Y0-BUB-344

SCIENCE, THE SINGULAR, AND THE QUESTION OF THEOLOGY

THE NEW MIDDLE AGES

BONNIE WHEELER, *Series Editor*

The New Middle Ages presents transdisciplinary studies of medieval cultures. It includes both scholarly monographs and essay collections.

PUBLISHED BY PALGRAVE:

Women in the Medieval Islamic World: Power, Patronage, and Piety
  edited by Gavin R. G. Hambly

The Ethics of Nature in the Middle Ages: On Boccaccio's Poetaphysics
  by Gregory B. Stone

Presence and Presentation: Women in the Chinese Literati Tradition
  by Sherry J. Mou

The Lost Love Letters of Heloise and Abelard: Perceptions of Dialogue in Twelfth-Century France
  by Constant J. Mews

Understanding Scholastic Thought with Foucault
  by Philipp W. Rosemann

For Her Good Estate: The Life of Elizabeth de Burgh
  by Frances Underhill

Constructions of Widowhood and Virginity in the Middle Ages
  edited by Cindy L. Carlson and Angela Jane Weisl

Motherhood and Mothering in Anglo-Saxon England
  by Mary Dockray-Miller

Listening to Heloise: The Voice of a Twelfth-Century Woman
  edited by Bonnie Wheeler

The Postcolonial Middle Ages
  edited by Jeffrey Jerome Cohen

Chaucer's Pardoner and Gender Theory: Bodies of Discourse
  by Robert S. Sturges

Crossing the Bridge: Comparative Essays on Medieval European and Heian Japanese Women Writers
  edited by Barbara Stevenson and Cynthia Ho

Engaging Words: The Culture of Reading in the Later Middle Ages
  by Laurel Amtower

Robes and Honor: The Medieval World of Investiture
  edited by Stewart Gordon

Representing Rape in Medieval and Early Modern Literature
  edited by Elizabeth Robertson and Christine M. Rose

Same Sex Love and Desire Among Women in the Middle Ages
  edited by Francesca Canadé Sautman and Pamela Sheingorn

Listen Daughter: The Speculum Virginum and the
Formation of Religious Women in the Middle Ages
  edited by Constant J. Mews

SCIENCE, THE SINGULAR, AND THE QUESTION OF THEOLOGY

Richard A. Lee, Jr.

palgrave

First published 2002 by PALGRAVE™
175 Fifth Avenue, New York, N.Y.10010 and
Houndmills, Basingstoke, Hampshire RG21 6XS.
Companies and representatives throughout the world.

PALGRAVE is the new global publishing imprint of St. Martin's Press
LLC Scholarly and Reference Division and Palgrave Publishers Ltd
(formerly Macmillan Press Ltd).

ISBN 0-312-29296-1

Library of Congress Cataloging-in-Publication Data
Lee, Richard A., Jr.
Science, the singular, and the question of theology / Richard A. Lee, Jr.
    p. cm.—(The new Middle Ages series)
  Includes bibliographical references and index.
  ISBN 0-312-29296-1
  1. Theology, Doctrinal—History—Middle Ages, 600–1500. Philosophy,
Medieval. I. Title. II. Series.

BT26.L44 2002
23'.2'0902—dc21

                                  2001035958

A catalogue record for this book is available from the British Library.

Design by Letra Libre, Inc.

First edition: January 2002
10 9 8 7 6 5 4 3 2 1

Printed in the United States of America.

CONTENTS

To all my teachers,
especially Fernenc Feher
and Reiner Schürman,
who could not live to see it.

PREFACE

The argument presented here appears, on its face, to be a relatively straightforward historical presentation of what happened to the question of the scientific nature of theology from Grosseteste to Ockham and his reception. Yet the argument is, in fact, constructed in reverse. Rather than attempting to tell the story as if we did not know the conclusion, the story is told precisely from the point of view of the conclusion. How can we read the history of medieval thought (especially on the question of the scientific status of theology) such that Ockham's position appears to be a "natural" one? How do thinkers such as Aquinas and Scotus appear if we know already the problems and issues with which Ockham deals? Can we read the history of medieval thought from the perspective of Ockham's thought?

What motivates the choice of such a form of history is a relatively simple issue. It has always seemed strange to me that from a certain moment in history, that is, from the moment of the condemnations of the 1270s, Aquinas could be read as a radically dangerous thinker and Ockham as the answer to that threat. To be sure, there was a certain rivalry between the mendicant orders. To be sure, Bonaventure seemed to be moving theology along a different path than Albert the Great or Aquinas. But such rivalries and traditional allegiances cannot fully explain how the Franciscans seemed to emerge victorious from the various condemnations of the 1270s. The history presented here attempts to be a history of how Ockham's "nominalism" was for a time preferable to Aquinas's "realism." That is, this history stops at Ockham and his reception and attempts not to rush headlong into the counter-Reformation where Aquinas would once again prove himself the better. I make no claims to offering a better understanding of what Grosseteste, Aquinas, Scotus, or Ockham "really meant." Rather, I simply try to show a way of reading a particular history that makes sense of the contributions of all thinkers from within that history. I attempt to read the history of this question such that Ockham appears not as a destructive radical, but rather as attempting to solve problems raised by an Aquinas who can appear to be radical. What is opened in this form of telling the story is a reading of Aquinas that no longer seems possible to us today. But unless we can reopen this way of reading Aquinas, we run the risk of misunderstanding later medieval responses to Aquinas.

This points to another reason for offering a history such as this. The history of reading medieval philosophy after the Middle Ages is itself an interesting topic for scholarly study. What are the effects of our interpretations of medieval thought in the aftermath of the counter-Reformation? The fact that Aquinas was the philosopher to whom the Catholic Church turned for a response to Luther meant that Aquinas had to be read in a different way. But for the same reason, thinkers like Scotus, Ockham, Marsilius of Inghen, Pierre d'Ailly, Robert Holcot, and others had to be reinterpreted with Aquinas serving as the standard. Aquinas's historical position was thereby reversed; he now comes *after* Ockham and is a response to Ockham. Because of this peculiar history, Ockham's reading of and response to Aquinas cannot be detached from the victory that Aquinas won over Luther.

This study, therefore, attempts to uncover just how medieval Aristotelian thought from before the condemnations of the 1270s might have been interpreted in the aftermath of the condemnations. I have had to avoid raising debates in Aquinas scholarship that have become standard. I do not mean to say that such debates have no merit, that such debates fundamentally misunderstand Aquinas. Rather, I mean only to say that I am interested in how philosophers, particularly Franciscan philosophers, after 1277 could have understood and misunderstood Aquinas.

The choice of presenting the story in this fashion leads to a certain way of dealing with the scholarly contributions of those giants on whose shoulders I sit. I have made a deliberate choice to attempt to refrain from pointing to the scholarly literature when I am in agreement or disagreement. I have made this choice for two main reasons. First, I did not want my quibbles with others' interpretations to cloud the argument I am pursuing unless I find that the disagreement *fundamentally* affects that argument. Second, I have traced my indebtedness to previous scholars in the bibliography and one who wishes to know the exact extent of my agreement or disagreement can certainly trace that by reading the texts referred to there. I do this more out of honor for the work of these scholars than out of the vanity of thinking that my contribution is better than theirs.

I should add, perhaps, one final note of caution. I have come to the study of medieval philosophy in general and this topic in particular from an interest in contemporary "continental" philosophy. This is unusual today, but it was not always so. Hegel, Brentano, Heidegger, Arendt, and many other of the major figures in this tradition were constantly engaging medieval thought. Yet today the tradition of continental philosophy and the tradition of scholarship into medieval philosophy have parted ways, neither showing particular interest in the other. Perhaps in some small way this study can be a move toward bringing these two sides of the family back into harmony.

INTRODUCTION

THEOLOGY, SCIENCE, AND RATIONAL GROUND

The Concept of Rational Ground

From its very beginning, philosophy set for itself the challenge of pursuing specific sorts of answers to a seemingly trivial question: Why? This question admits of a wide range of answers, it seems. Yet in its infancy, philosophy set about putting this question with a certain kind of precision. Religion, myth, and tragedy all ask this kind of question and all give answers of specific kinds to this question. Philosophy, at its inception, distinguished itself from these by defining what the question means and what would count as an answer.

The question specific to philosophy is a combination of epistemology and metaphysics. Philosophy, already in Plato, understands that the answer to the question must fulfill two conditions: (1) The answer must explain how we come *to know* that about which we are asking the question; and (2) the answer must explain how that about which we are asking the question came *to be*. In the *Phaedo,* Socrates puts this succinctly: "It seemed to me splendid to know the reasons for each thing, why each thing comes to be, why it perishes, and why it exists."[1] This precise question, that is, the question of the reasons of the thing that perform both an epistemological function and a metaphysical function I call the question of the rational ground.[2]

The rational ground, therefore, is a specific kind of answer to the question "Why?" For it is an answer that appeals to reason (epistemological) and holds that, in one way or another, reason is the cause of a thing (ontological). In short, the rational ground is an answer that attempts to join the way in which a thing comes to be with the way in which a thing is known. The shortest expression of the rational ground comes from Aristotle's *Posterior Analytics:* "We think that we know each [thing] . . . when we think that we know the cause through which the thing exists . . ." (71b10).

Rational ground joins the order of genesis (coming to be) and the order of knowledge (coming to know).

The rational ground provides a kind of answer that is different from, for example, the answer of a myth or a story of creation. For a story of creation provides a ground for the coming to be, passing away, and existence of a thing, but it does not provide a *rational* ground. A story of creation merely posits that the existence of a thing is the result of some kind of divine creative activity that is itself not further explicable in terms of its causes. There is no *rational* ground where the ground offered is not subject to reason. One can read the section of the *Phaedo* cited above as Plato's attempt to distinguish the kinds of answers that are properly philosophical from all other kinds of answers. Philosophy is the search for the *rational ground,* and this has certain consequences.

The first consequence of the search for rational ground is that the ground, that is, that which has metaphysical responsibility for a thing coming to be or existing, is *rational,* that is, it appeals to reason. When Socrates finds the writings of Anaxagoras, he is pleased because it seemed to Socrates that "intelligence (*nous*) should be the reason for everything" (97c4). Socrates finds that Anaxagoras, in the end, supposes that things like air and ether and water are the causes.[3] "But to call such things 'reasons' is absurd" (99a5). Anaxagoras's primary mistake was to offer as a ground that which is not rational. Air, ether, and water cannot be the rational ground because they are not, according to Socrates, subject to reason.

This leads to a second consequence. If the ground, that is, that which is responsible for the existence and coming to be, of a thing is to be a *rational* one, then it will be difficult to separate this ground from reason itself. This again is seen clearly in the story that Socrates tells of Anaxagoras. For what appealed to him in Anaxagoras was the fact that reason, intelligence, *nous,* was posited as the cause for all things. This is a direct result of the ground's being *rational.* For in the very notion of rational ground, as we have seen, the reason for something coming to be is precisely the reason for our being able to know the thing. Thus, the ground of the coming to be or existence of the thing is not only a metaphysical ground, but an epistemological ground. In this way, the ground is often reason itself.[4]

This leads to a final consequence. Since the attempt at uncovering the rational ground of something existing or coming to be is both metaphysical and epistemological, it is also general in a certain sense. There is not an infinite plurality of rational grounds, one for each singular. Rather, the rational ground itself is an attempt to explain, *on another level,* why something exists in the way it does. This requires, therefore, that the reason for existing be more general than the singular that is existing before us. When we ask why a human is a human, or even why *this* human is human, we ask not about the singular ground that is valid only for this individual. Rather, we ask for a ground that is the same for all beings of this kind, or even for all beings in general.

This is one of the conditions that makes the ground rational. For if each thing required a singular and individual answer to the question "Why," then the ground would not be rational but would rather look something like a simple "because." If the answer to each "why?" question were singular and individual, then all answers would cease to serve their epistemological function: "This human is human because it is." Therefore, in moving from the singular to its rational ground, the singular is no longer important—the rational ground is ground for *any* singular whatsoever. The disappearance of the singular before its rational ground, however, leads to a peculiar problem with the search for the rational ground itself.

The Singular and Its Rational Ground

We will see in the next chapter how Aristotle's *Posterior Analytics* situates itself within the problematic of the relation between an existing singular and its rational ground. Here, however, we can uncover the general contours of the problem. The search for rational ground always begins with something whose ground we are seeking. This is a singular that is found to be existing before us. It is this thing whose ground we are seeking. We want to know what it is and why it is, that is, we look for the "what it is" in such a way that the "what" answers the question "why." Without this existing singular being there before us, the search for the rational ground would never get underway.

Yet once we have determined that the question "why" must be answered by way of the rational ground, we already admit something prior (ontologically) to the existing singular. What is here before us is the existing singular, and this has epistemological priority in that it is the first that is grasped. However, once the rational ground is posited, the rational ground is seen to be that through which the existing singular is truly known because the rational ground is seen as the cause. We can say that we know the singular, but it is known truly only when its rational ground is known.

The rational ground as explanation fulfills both an ontological and an epistemological role. Because of this, we have seen that the rational ground requires a certain level of generality. The rational ground, however, is supposed to be the ground of an existing singular. There arises now a problem with the relation between an existing singular and its rational ground—a ground that must be more general than the singular. Therefore, what was supposed to serve as an explanation for the existence or coming to be of a singular thing does not itself have singularity. The rational ground can serve as an explanation for existence or coming to be, but not *this* existence, *this* coming to be. The rational ground cannot account for the singularity of the singular.[5]

There is a trade-off, then, between the search for the rational ground and grasping the existence of the singular. For Plato, the issue for philosophy was that it should provide causes (reasons) why things are, come to be, and pass away. The passage of the *Phaedo* under consideration here makes clear that philosophy, in providing answers of this kind, distinguishes itself from myth—which also provides answers to these questions. Yet the answers that myth offers usually have to do with the powers of the gods, which are not understandable through reason—they have no rational ground.[6] What the mythic (and religious) accounts of creation provide is not the reason, but the power that brings things into existence. Because of this, myth is always about the singular, to some extent.[7] In pursuing the rational ground as the answer to these questions, philosophy certainly makes the cosmos more intelligible, but does so only by losing the ability to understand the existing thing in its singularity.

Philosophy distinguishes itself from myth only because myth never presupposed a rational design to the universe, that is, myth never supposed that the creator(s) of the universe create(s) the universe so that it be rationally grasped. What happens, however, when the creative God of Genesis is thought to have created a world that is understandable through rational grounds? In short, what happens when Christian theology brings together the power of creation with the rationality of ground?

Creative Power, Rational Ground, Existing Singular

Both Plato and Aristotle try to set forth the conditions that would make an account rational. Myth answers the question about the coming to be, existence, and passing away of things by turning to the power of the gods.[8] The story of creation in Genesis refers also to divine power that is not subject to further accounting. Tragedy refers to the relation between fate and power. These three types of answers, in short, are not so much about giving an account of the coming to be, existence, and passing away of things as they are about how the world is not ours—it is not of our making and it is not of our possession. Myth and tragedy tell us that the things about which we ask these questions are "without why."[9] Since things are without reason, all we are left with are our lives among existing singulars, be they inanimate things, human beings, or gods. Because myth and tragedy refer only to divine power and *not* to divine reason, they do not and cannot provide the ground for existing singulars—or, rather, they provide a "groundless ground," that is, a ground that points only to divine power that is itself without ground.[10] In referring to divine power, myth and tragedy refer to something that does not admit of generality. Zeus swallowed his children, but that is not the ground of cosmos in general.

Plato and Aristotle, conversely, set forth certain conditions for giving an account such that there is no ground that is not grounded. Nothing is "without why." If they are to refer to divine power in creation, then that power itself must have a ground. This is the force of Socrates's statement that if Anaxagoras is right, then each thing comes to be, exists, and passes away because it is "the best," or "for the best." Once we grasp this, the existing singular is provided a reason. This is also the force of Aristotle's refusal to entertain creation at all. A world that is eternal is one without a ground in divine power and, therefore, one whose ground can be grasped.

When the question of the rational ground as posed in Aristotle's *Posterior Analytics* is forced into confrontation with the creative power of God in the form of the question "whether theology is a science," the question must, as we have seen, focus on *both* the existing singular *and* the rationality of God's creative power. In other words, the question of whether theology is a science is the question of whether God's creative power is a groundless ground and, consequently, whether existing singulars too appear as without ground. Traditionally this question is posed under the debate between realism and nominalism. However, the issue is not merely the ontological status of universals, but is the more serious and penetrating question about the rational intelligibility of God's creative activity. It is the question of whether the rational ground can ever be adequate to that for which it is ground.

The Plan of the Work

In what follows, the question of the relation of the existing singular to its rational ground will be the focus of the investigation into the medieval question of the scientific status of theology. I begin with an investigation into Aristotle's *Posterior Analytics,* which provides the definition and methodology of science [*episteme, scientia*] that informs the medieval debate. I uncover the fundamental doctrine that will concern our investigation: how we move from sensation of singulars to universal propositions. I then show how this doctrine raises problematic issues if one considers theology as a science.

I then investigate Robert Grosseteste's interpretation of *Posterior Analytics* because it was one of the earliest and most influential commentaries on Aristotle's text. This reading will set the stage for our understanding of medieval interpretations of *Posterior Analytics*. After these two preliminary investigations, I trace the history of the question of the scientific status of theology through Aquinas, Duns Scotus, and William of Ockham, and attempt to show how the problematic of the relation of existing singulars to their rational grounds appears in each of these attempts to answer the question of theology as science. Finally, I will indicate some later medieval ways (Marsilius of Inghen and Pierre d'Ailly) of taking up the issue of the rational ground in its relation to the existing singular.

My argument is that in the later Middle Ages, the search for the rational ground was forced into conflict with God's creative power. This conflict arose, mostly, out of theological concerns (especially as is evidenced in the Condemnations of 1277). However, the consequence of this conflict was that existing singulars emerge without ground. This was the main task that philosophy had to confront.

CHAPTER 1

THE CONTACT OF SCIENCE AND THEOLOGY

The Origin of the Question

Medieval philosophy did not always have the tools by which the question of the scientific nature of theology could be asked. We can date the question with relative specificity: The question cannot be asked before about the year 1156.[1] For the question of science is a question raised in Aristotle's *Posterior Analytics* and this text does not make its appearance in the Latin West until sometime around that year. However, in terms of its appearance in philosophical and theological texts, let alone in terms of commentaries of it, *Posterior Analytics* does not become an influential text until the early thirteenth century. It is only at that point, then, that the very question "Is Theology a Science?" can be asked.

The *Posterior Analytics* remains, in many ways, one of Aristotle's most enigmatic texts. There is still debate, in fact, about just what this text is supposed to do. Is it a text that teaches the scientist how to pursue research? Or is it a text that merely shows the scientist how best to organize and present research?[2] Even without attempting to answer the question of the goal of the text, commentators are still lacking consensus on what its more important doctrines are and how they function. Adding to this puzzle is the fact that Aristotle himself never seems to have put to use the methods set forth in *Posterior Analytics*.[3]

In its bare outlines, the *Posterior Analytics* presents an analysis of a disposition (gr. *hexis,* lat. *habitus*) that he calls *episteme.* We have this particular cognitive disposition of syllogisms only of a particular kind—those that are demonstrative. The text, then, spends most of its energy on detailing just what kind of syllogism counts as demonstrative—that is, it lays out the rules that a syllogism must obey in order to be counted as demonstrative and thus be available to the cognitive disposition *episteme.*

The Premises of a Scientific Syllogism

Because *Posterior Analytics* is concerned with a disposition that relates to a particular kind of syllogism, it presupposes the syllogistic logic treated in *Prior Analytics*.[4] It will not be concerned, then, with the basic theory of the syllogism, rules for the validity of a syllogism, conversion of various forms of syllogism, and so on. What it will be concerned with, however, is just what requirements need to be fulfilled in order for a syllogism to count as demonstrative. These requirements will fall into two basic kinds: (1) requirements as to syllogistic form and (2) requirements as to syllogistic content. The requirements of form are relatively moderate. Aristotle holds that a demonstrative syllogism is most properly that syllogism in the first mood of the first figure—that is, "BARBARA."[5]

The requirements as to content are what take up the major portion of *Posterior Analytics*. Content, here, refers not to the particular terms that make up the propositions of a syllogism. Rather, content concerns the *kind* of premises that are required in a syllogism that is demonstrative. The premises must be "true and primitive (*protos*) and immediate (*amesos*) and more known (*gnorimos*) than and prior to (*proteros*) and explanatory (*aitios*) of the conclusion."[6] Furthermore, Aristotle tells us that these premises must be necessary, known in themselves [*kath' auto*] and universal [*katholou*].[7]

These requirements may be related to one another (as, for example, necessity is related to "in itself") or they may be radically different (as, for example, true and explanatory). Taken together, however, they show that whatever a demonstrative syllogism is, it requires premises that themselves are not demonstrated. The indemonstrability of premises is required for Aristotle because he wants to show that we know something, but that knowledge is not from recollection. The indemonstrability of the premises is Aristotle's solution to the Meno Paradox. Let us look briefly at each of these requirements.

- *True.* It is not the case that the premises of all syllogisms must be true in order to produce a true conclusion. However, since a demonstrative syllogism is to expose the *causes* of the conclusion or the *reason why* the conclusion is true, such a syllogism needs to have true premises. It is the causal relation, therefore, between a conclusion and its demonstration that leads to the requirement that the premises must be true.
- *Primitive.* According to Barnes, "primitive" means that there is nothing from which the proposition in question can be derived.[8] It is for him, as it seems to be for Aristotle, related to "non-demonstrable." This means that the predicate must belong to the subject qua itself

and not qua some third thing. Barnes, however, analyzes "primitive" in such a way that it becomes almost indistinguishable from "immediate." Aristotle seems to be indicating by "primitive" that the predicate belongs essentially to its subject. Here "essential" would pertain to what a thing is and "primitive" would indicate that the predicate does not belong to any other subject of another species of the same genus.

- *Immediate. Amesos* means to lack or be without a middle term. In a demonstrative syllogism, the question of the middle term is crucial. At 90a6, Aristotle claims that the middle term *is* the explanation. Therefore, since there can be no demonstration of such premises, they must lack a middle and are, consequently, indemonstrable.[9]

- *Prior to and More Known.* Barnes suggests that whereas Aristotle often distinguishes between priority and knowledge (or knowability), here he identifies the two.[10] But Aristotle seems to consider this to be one requirement, not two separate things. The type of priority Aristotle has in mind here is precisely an epistemic priority. What is difficult, however, is that what is epistemically prior is actually temporally posterior in our coming to know it. Aristotle raises a familiar distinction between what is prior in relation to us and what is prior without qualification (or according to nature): "I call what is prior and more known in relation to us what is nearer to perception, prior and more familiar *simpliciter* what is further away. What is most universal is furthest away, and the particulars are the nearest; and these are opposite to each other" (71b35–72a5). The universal and the cause of something are prior to, even in an epistemic sense, the singular and that which is caused. Temporally we come to know what is closer to us first. This temporal priority does not contradict the fact that the universal and the cause are, once known, epistemologically prior to the singular and the effect.

- *Explanatory of the conclusion.* This follows directly from Aristotle's definition of the kind of knowledge that a demonstrative syllogism brings. Science is knowledge of the causes of some fact. Therefore, the premises used in the syllogism demonstrating that conclusion must contain the causes, and consequently the explanation, of the conclusion. As Barnes says, "If the only knowledge necessary for having a demonstration of *P* is knowledge of the principles from which *P* is deducible, then the principles must contain the explanation of *P*."[11]

- *Necessity, In Itself, and Universality.* These requirements arise because of the nature of the specific kind of knowledge Aristotle is investigating: "Since it is impossible for that of which there is understanding *simpliciter* to be otherwise than it is, what is understandable in virtue of demonstrative understanding will be necessary" (73a 21ff.). Aristotle argues that if the conclusion of a demonstrative syllogism must be necessary then the premises must be necessary. This is not because a necessary conclusion follows *only* from necessary premises. Rather, it is because the premises are to show the reason for the conclusion. Thus, if the conclusion is necessary the premises must also show why. The premises must be necessary because they are explanatory of a necessary conclusion.

The other two conditions, that is, that the premise be universal and be a predication "*in itself,*" seem to be nothing other than an explication of the type of necessity Aristotle has in mind. For he says that "whatever is universal belongs from necessity to its objects" (73b29) and that universal is "whatever belongs [to something] both of every case and in itself and as such" (73b25–26). Furthermore, the entire discussion of "belonging to every case" and "universally" and "in itself" is introduced immediately after Aristotle raises the issue of necessity: "Demonstration, therefore, is deduction from what is necessary. We must therefore grasp on what things and what sort of things demonstrations depend. And first let us define what we mean by [holding] in every case and what by in itself and what by universally" (73a24ff.). Just what Aristotle means by "in itself" will be one of the main problems of the medieval debate on the scientific nature of theology. How a thinker interprets this condition will affect how that thinker understands whether or not theology is a science. Therefore, a discussion of the meaning of this condition must be postponed for now and will be taken up over and over again throughout this investigation.

Grasping the Premises of a Scientific Syllogism

We have already seen that one of Aristotle's main concerns is to ensure that there is no infinite regress in our scientific knowledge. This concern led him to posit the requirements listed above, which the premises must fulfill in order for the syllogism to be considered "scientific." However, there is one crucial issue that has not yet been addressed and that has the capacity to destroy Aristotle's entire project. How is it that we come to have knowledge of the premises of a scientific syllogism if not through demonstration? If Aristotle cannot offer a way of grasping the premises such that they are known and fulfill all the requirements he has set forth, then the entire notion of a demonstrative syllogism is for naught.

Aristotle is remarkably silent on this issue. Within *Posterior Analytics,* this subject occupies only one short chapter located at the end of the second book. This chapter (II, 19), which purports to clear up how the principles of a demonstration become known and what the state of the soul is that becomes familiar with them, continues to perplex commentators on *Posterior Analytics*. However, this is the main issue that will confront anyone concerned with whether or not a specific body of knowledge is "scientific." If it can be shown that the first principles of a given body of knowledge cannot, by their very nature, become known at all in the way Aristotle says they must, then that body of knowledge cannot be properly called "scientific."

Aristotle's own remarks on the subject seem to lead to more problems than they solve. His remarks can be divided into two main parts: (1) those dealing with the cognitive disposition (*hexis*) that grasps the principles and (2) the way in which the soul goes about grasping those principles. As far as the cognitive disposition is concerned, Aristotle clearly thinks that *nous* is just such a disposition. While *nous* may have other tasks it can perform, at least one of its tasks is to grasp first principles.[12]

As far as how this grasping takes place, there seems to be no clear consensus. Aristotle begins his treatment of the grasping of first principles by distinguishing among various types of animals based on certain capacities. First, there is that which is common to all animals, namely, perception. Some animals, however, have the further capacity for retaining what has been perceived. Among those animals with this capacity for retention, there are those that come to have an account (or formula) from the retention of such perceptions. Put in another way:

> So from sense perception there comes memory, as we call it, and from memory (when it occurs often in connection with the same thing), experience; for memories that are many in number form a single experience. And from experience, or from the whole universal that has come to rest in the mind (the one apart from the many, whatever is one and the same in all those things) [there comes] a principle of art and of science—of art if it deals with how things come about, of science if it deals with what is the case (100a4–10).[13]

The text may have many ambiguities, but its overall contour seems clear. There is a movement from sense experience of a singular to the grasping in the mind of a principle. This principle seems to be based on some sort of universality, which is the middle ground between the particular and the principle. The principle, then, will itself have universality. The principle will be something of the form All X is Y—it is a "connection."

> The preceding discussion has introduced three conceptions of universals: universal concepts (horse), universal connections (all triangles have 2R), and universals as grounds of conclusions. These conceptions are interrelated: a

universal connection is a connection between an attribute and a universal concept, and the universal ground of a conclusion is a universal connection.[14]

The process by which all of this happens is called "induction" (*epagoge*). This process seems to entail the ability of the mind to draw out from many experiences of singulars that which is "one and the same in all those things." When this happens, a universal "makes a stand" in the mind. It is from these universals, then, that the principles are derived. Aristotle seems to have made good on his promise of science: It avoids an infinite regress because the principles of science are not demonstrated but arise out of perception.

Just precisely *how* this whole process happens is still, however, somewhat of a mystery. Aristotle has given us the stages, but not the means to move between the stages. More seriously, there seems to be a sort of circularity in Aristotle's account. McKirahan, for example, tells us, "*Epagoge* is the way we come to spot individuals as individuals of a kind. Equivalently, it is the way we become aware of universals in particulars."[15] Yet induction cannot be both at the same time. For to spot individuals as being members of a kind presupposes that we already know the kind. Conversely, "to become aware of universals in particulars" leaves aside the account of what it is in the singulars of which we become aware in induction.

This is the heart of the problematic that will occupy medieval thinkers in their attempt to determine whether theology is a science. Yet the issue also goes to the very heart of the philosophical enterprise. The entire logico-epistemological edifice that Aristotle has built now depends upon the ability to bring existing singulars into its structure. The logic of *episteme* itself is not discovered in the world in which one already finds oneself. However, it is precisely that world that ought to give itself up to an account in *episteme*. In a sense, then, *episteme* is always chasing after the existing singulars. While *episteme* appears to be *posterior* to the existing singulars, Aristotle argues that it is able to discover what is ontologically and, therefore epistemologically *prior* to such things. The entire logico-epistemological framework of causality is what allows Aristotle to move behind the sheer givenness of singulars and attempt to explicate their "reasons" for being. Aristotle's theory attempts to situate itself between two extreme solutions to this problem. On the one hand, he refuses to give up the possibility of *episteme*, that is, of giving a rational account of existing singulars. Therefore, he wants to maintain that we can grasp the causes of things. On the other hand, he refuses to give up the existing (sensible) singulars, that is, he refuses to collapse the being of these singulars into knowing reason. The singulars are neither the result of knowledge nor do they stand completely outside knowledge. Such a middle terrain, however, is not easily occupied.

It is not necessary that we solve all the riddles of this short section of *Posterior Analytics*. For this will be the task of any philosopher interested in showing that theology is a science. It is because of Aristotle's requirements pertaining to the premises as well as his introduction of the way in which we grasp the principles of a science that the very status of theology as a science was called into question. We need, then, to investigate just how these issues relate to the question of theology.

Theological Propositions and Aristotelian Science

If theology is to be a science, then it must fulfill the requirements Aristotle sets forth in *Posterior Analytics*. The syllogistic requirements present little or no problem for theology. Certainly theological propositions can be arranged in syllogistic form as much as any other kind of propositions. The main problem philosophers and theologians had with this issue is that it seems as if theology cannot fulfill the requirements that pertain to the content, that is, to the kinds of propositions that can be premises in a demonstrative syllogism.

Whatever else it may be and do, theology, if it would be a science, would have God—or some aspect of the divine—as its subject. This means that theology would begin with self-evident (*per se*) premises that pertain to God. According to *Posterior Analytics,* such premises would include definitions as well as predications of properties that belong essentially to their subjects. Furthermore, these propositions would not be demonstrable and, therefore, would be grasped by *intellectus* (i.e., by *nous*).

Intellectus would be the disposition (*hexis, habitus*) that gives insight into the essence of that subject such that the predicate is seen to belong to it. Is such insight possible when God is the subject? This is precisely the main issue upon which the question of the scientific nature of theology was seen to rest. If *intellectus* of God's essence is not possible, then theology cannot be a science. For if we do not grasp the divine essence, the propositions that would form the scientific, theological syllogism would not fulfill the conditions such premises must have according to Aristotle.

Yet all later medieval thinkers agreed that *intellectus* of God is possible—there can be a grasping of God's essence. For this is precisely what the "beatific vision" is. The blessed enjoy seeing the divine essence, and this "seeing" is most often described as precisely the kind of grasp one has of self-evident, primary, and immediate truths. Thus, in general the grasping of the divine essence that would be necessary if theology is to be a science is possible. Is it, however, possible in this life, that is, before beatitude?[16]

While we will see a range of answers to this question, most thinkers will answer that *intellectus* of the divine in this life would violate our state as "wayfarers" (i.e., those between the Fall of Adam and Eve and the state of blessedness) and is, therefore, impossible. This means either that one must assert that theology is not a science or one must develop a new account of how the requisite grasp of premises can happen outside *intellectus*. What will remain constant, however, is that the question of *intellectus*, and its relation to premises that are self-evident (*per se*), is *the central question* of the scientific nature of theology.

When the ambiguity of Aristotle's *nous* is brought into contact with theology, the issues and problems raised by Aristotle become all the more urgent. For the question of the scientific status of theology raises the issue of bringing *episteme*, that is, the knowledge of the reasons, into the divine essence itself. As long as the divine essence resists being brought into the realm of rationality, the ultimate causes of all things also will stand outside the grasp of knowledge. Because Aristotle's theory is one in which a thing is known when its causes are known, the question of the application of this theory to the divine is the question of grasping the causes of things in general, or of grasping the ultimate cause of things.

Most medieval theologians agreed that God's essence coincides with God's existence.[17] If *intellectus* of God would be possible, it would mean that an immediate grasp of an existing singular, precisely as existing singular, would be possible. Such a grasp would allow rationality to trace the cause of existence of mundane singulars to their foundation in the existence that belongs to God. This extension of rationality, however, requires that divine causality be brought under the gaze of rationality as well. In short, the fact that God's essence is not able to be the object of *intellectus* means that a crucial link between God and the world is severed—divine causation in creation is not subject to *episteme*.

If, however, creation itself stands outside of reason's grasp, then it seems that reason is also unable to grasp scientifically the created things. For the knowledge of causes that Aristotle posited as the crucial element of *episteme* will always founder on the ultimate contingency of divine creation. The question of the scientific nature of theology, therefore, is the question of the rationality of the world in which we already find ourselves. The contest is between reasserting that rationality above the insistence that God is radically unknowable and insisting on the utter contingency of the created world even if that means, in the end, that it falls outside the grasp of reason.

The arguments carried on in the Middle Ages are decidedly theological. That is to say, one does not see medieval thinkers championing the radical contingency of existing singulars as a value in itself. Rather, the fact that existing singulars always exceed rationality is a result of the theological insistence that God's creation is dependent only on the divine will. In many ways, the insistence on the extra-rational character of the divine will is a conservative (even reactionary) theological position. Its philosophical consequences, however, are quite revolutionary. Once the divine is placed outside of *episteme,* causality itself ceases to provide the metaphysical basis of rationality. The existing singular, qua singular, becomes a matter of intense focus. For a moment in the history of medieval philosophy, rationality is jarred in the face of the excessive nature of existence. This moment proved decisive. For this moment is one of crisis—a crisis of the foundations of reason. For when the objects can no longer provide the basis for reason, philosophy can turn toward the subject. However, this is only one possible solution to the crisis. And if we are to critique that solution, then we must stand for a moment at the crisis to see that this particular solution was not inevitable.

CHAPTER 2

DIVINE IDEAS, ARISTOTELIAN SCIENCE: ROBERT GROSSETESTE AND THE THEORY OF *SCIENTIA*

Introduction

Grosseteste was perhaps the first medieval Latin-speaker to comment extensively on *Posterior Analytics*. There are references, however, to *Posterior Analytics* in the period immediately before Grosseteste, though these references show no extensive use, familiarity, or grasp of the text.[1] Grosseteste himself was an incredibly productive translator and commentator, addressing himself extensively to the Bible, the works of Pseudo-Dionysius, as well as several texts of Aristotle—most notably the *Ethics*. However, it is because Grosseteste became the first lecturer to the Franciscans at Oxford (as well as Oxford's first chancellor) that his influence was felt as deeply and as long as it was.

Grosseteste was heavily influenced by a Neoplatonic tradition, as is evidenced by his interest in Pseudo-Dionysius. He was also a man of science. His commentary on *Posterior Analytics* reveals a great deal of Platonic and Neoplatonic concepts: He relies upon Platonic forms, he stresses a metaphysics of light, he holds a doctrine of divine illumination. His commentary, however, set the terms of the debate in many ways and thus it is fitting that we turn to his text and his interpretation of (1) self-evident propositions and (2) the cognitive disposition of *intellectus*.

Per-se Predication

Grosseteste begins his analysis of what it means "to say" something "*per se,*" with a discussion of what it means *to be per se*.[2] There are three ways in which something can exist "through itself." The first is when something does not exist through an efficient cause. In this sense, something exists through itself because it is not caused by any other thing. According to this sense of "*per se*," only the first cause will be *per se,* and all other beings will exist through some cause. Therefore, this first way of understanding *per se* is as non- or self-caused.[3]

Secondly, *per se* is said of that which does not exist through a material cause. In this sense, only intelligences (i.e., the angels) are *per se.* This sense refers to those beings that exist not in some combination with matter, but on their own. Hence they exist through themselves, that is, through their own being understood as their form. This group, however, obviously contains beings that are not *per se* according to the first definition.

Lastly, those beings are *per se* that do not exist through a subject. That is, only substances exist *per se,* and all accidents would exist through their subjects. All three of these ways of understanding "*per se*" refer to beings that somehow exist through themselves, that is, they need, in one sense or another, no other being for their existence. One could say that all these types of *per se* beings are *in themselves per se,* or existing *per se* because of the kind of beings they are. On the other hand, the kind of *per se* we are dealing with in a demonstrative syllogism is per-se *predication.* In predication, two things are necessarily involved: the subject and the predicate. Therefore, per-se predication cannot refer to per-se existence because such perseity refers only to one thing as being *per se* in itself. Per-se predication requires a *relation* between things or terms. Thus it is not the *beings* to which "*per se*" refers but their relation. The type of *per se* we need to investigate is "*per se alterum de altero*" (111). This type of *per se* obviously has different characteristics than a being that exists *per se.* There are many types of per-se predication, according to Grosseteste, yet there are only two that are relevant to demonstration.

Generally speaking, one thing will be said of another "*per se*" when the quiddity of the one essentially—that is to say, not accidentally—comes out of or "climbs out of"[4] the quiddity of the other. This connection means that the quiddity of the one, which has come out of the quiddity of the other, gets its being from that from which it comes, as from its efficient, formal, material, or final cause.[5] However, this relation must be such that the definition of that which gets its being includes that from which it gets its being. "*Per se*" refers to the connection between the two such that one climbs out of the other. This relation or connection, however, is a metaphysical relation between essences, as we shall see below.

Definitionally, there will be per-se predication when the genus or the differentia is predicated of the species. For only in this case will the definition of the one include the other. Thus, for example, if "rational" or "animal" is predicated of "human," this predication will be "*per se.*" However, we cannot lose sight of the fact that we are operating here in the middle ground between logic and metaphysics, or, to borrow a later phrase, between terms and things. Grosseteste is explicit on this point: " . . . this is the first mode of being or of saying *per se* of one to another" [. . . est primus modus essendi vel dicendi per se alterum de altero].[6] Per-se predication, then, as explicated along the lines of *egredior,* is a relationship of *saying* that is grounded in a specific relationship of *being.* This kind of predication is the expression of a relationship between essences.

Included in this mode of predicating *per se* are those predications in which the cause, either efficient, material, formal, or final, is predicated of its effect. Grosseteste says that what defines the first mode of per-se predication is that the predicate is "such that it is received in the definition of the subject" (112). Is that the case, however, when a cause is predicated of its effect? Is "wood" received in the definition of "house"? The problem raised by the predication of a cause is that while the effect certainly receives its being from the cause, it does not seem apparent that it receives also its definition from the cause. Grosseteste seems to want to bring definition and being into close proximity such that a predication in this first mode of *per se* exhibits both an ontological and a logical relation. He tries to save the predication of causes—which surely fulfill the requirement that the subject receive its *being* from the predicate—by showing that the cause would be predicated "obliquely" (i.e., in a case other than the nominative) of the effect. For example, "Smoke is *from fire,*" predicates the cause in a way that differs from, for example, "Humans are rational." What Grosseteste seems to be after is the effect *qua effect* includes the cause in its definition. This is indicated by the fact that he prefers, in this section, the term "caused" [*causatum*] to "effect" [*effectus*].[7]

The second mode of saying something *per se* of another is to predicate those accidents of a subject out of whose quiddity they essentially arise. While we are here predicating accidents, the relationship between the accident and its subject is such that there can be necessary predication. In this mode of per-se predication fall also those propositions in which the effect is obliquely predicated of its cause. The difference between the first mode and the second mode is that, as already stated, in the first mode the predicate is included in the definition of the subject while in the second mode the subject is in the definition of the predicate. This would mean, consequently, that in the second mode the predicate receives its being from the subject.

Now in these two modes of per-se predication, Grosseteste provides two sorts of definitions. On the one hand, the being, that is, *esse,* of the one thing comes out of the very quiddity of the other. On the other hand, the one is contained in the definition of the other. Therefore, the necessity involved in per-se predication seems to be of two kinds. First there is a metaphysical necessity that is grounded in the relation of *egredior*. In a relation in which the very being of the one thing arises out of the essence of another, it will always be the case that the one can be predicated of the other. Thus the proposition that predicates the one of the other will also be necessary. As such, the necessity is grounded in a metaphysical relation, that is, a relation between essences, and this relation is expressed with the term *"egredior."*

Secondly, if the definition of one term includes the other, then again it will always be true to predicate part of the definition of the defined. This necessity, then, will be logical because there will be an identity of sorts between the subject term and the predicate term. If the definition of "human" is "rational animal," then it is necessarily the case that humans are animals. Up to this point, everything is obvious. However, once the relation between these two definitions of *per se* (i.e., per-se *being* and per-se *saying*) is investigated, the situation becomes more complex. For we should ask whether or not one is prior to the other. At first glance, this question is easily resolved. The definition is nothing other than the "formula of the essence" of something. If something comes out of the essence of another, then that relation will also be seen in the formulae that express the essence of each of them. The definition of something is indeed metaphysical at its basic level—it will always express an essence.[8] Therefore, the entire idea of per-se predication, for Grosseteste, must arise out of this relation of *egredior*. This means, as has been said, that being is prior to saying.

Egredior *as the Basis of Per-Se Predication*

The fact that this relation of "coming out of" is the basis for per-se predication is clearly indicated when Grosseteste immediately follows the discussion of per-se predication with his next conclusion from *Posterior Analytics:* "Demonstration is a syllogism from those things which inhere *per se*" (129).[9] The point here is obvious: A syllogism begins with things that inhere in one another *per se,* and from the fact that these things are so related, a conclusion is able to be drawn in which the predicate inheres in the subject *per se.* On the other hand, only those things inhere *per se* that "come out of" the essence of the other. The relationship between predication and inherence is central to Grosseteste's understanding of the scientific syllogism.

We can use a sample syllogism in order to illustrate this: 'All animals are breathing and all horses are animals; therefore, all horses are breathing'. In this syllogism, according to the requirement we are discussing, breathing must inhere in all animals *per se* and animality must inhere in all horses *per se.* Now it is *because* of this series of per-se inherence that breathing is seen to inhere *per se* in all horses.[10] In a sense, then, the conclusion is nothing other than the unfolding of certain elements inhering in the middle term. Once the middle term is "unpacked" the conclusion is seen also to be a case of per-se inherence. The conclusion itself can be said to inhere in the middle term. The relationship that obtains between two terms, which is called "inherence," is such that the interval between them has a zero value, that is, the space between them is filled. This cannot be explicated according to the use and placement of the terms alone because perseity has a meaning that is not simply logical. Grosseteste's analysis of per-se predication is based on a relationship that obtains between the *being* of the predicate and the *being* of the subject, as was shown above. Thus, for Grosseteste the basis of *scientia* is essentially a metaphysical one. *Per se* marks, for him, the metaphysics of essential inherence. This means, in the end, a relationship of inherence and a predication in which that inherence is "unpacked."

Inhering in and Coming out of

Whenever a predicate inheres in its subject necessarily, that is, *per se* according to the first mode, and not accidentally, it is necessary that the one comes out of the substance of the other.[11] The link between the two is so strong that the destruction of the substance of the one implies the destruction of the other. Necessity, then, means that the thing that "climbs out" of the substance of something depends on that substance for its being. The destruction of the subject entails the destruction of that which inheres in that subject. This shows a relation of "inside" and "outside." That which inheres in another cannot be separated from that other. That is to say, the very predication of something that inheres in that of which it is predicated is a sort of false operation. Nothing separates the subject from the predicate. In this way, Grosseteste brings together these two spatial metaphors: inhering in and climbing out. It is the very relation between these two that grounds the necessity of per-se predication and, consequently, the scientific syllogism as a whole. Per-se predication is an expression of the fact that the predicate, by the mere fact of predication, has climbed out of that in which it essentially inheres. We will see that universality will be involved in these kinds of predications. Therefore, we cannot further explicate the details of this dual relation until we have investigated the nature of universals.[12]

We have this dual action on the part of those things that are *per se* or are said *per se* of another. First, the one must inhere in the other, that is, it must be such that destroying the one, the other is destroyed. On the other hand, the one must come out of the other, that is, not be posited as existing separately but be found in the very essence of the other.[13] One could say that per-se predication is the drawing out of something that essentially inheres in another thing. It would seem that "*egredior*" is the correlate to inherence when that which inheres is predicated of that in which it inheres. To climb out of the essence of another is the nature of this type of predication.

A demonstrative syllogism will be analyzed according to this dual relationship of "finding inside," and "climbing out." Per-se predication is the artificial separating of that which actually is found within the essence of that of which it is predicated. This is called "*egredior.*" The very artificiality of this separation and subsequent predication is a sign of its perseity. That which ought to be inside is found to have climbed outside. But the reason for this climbing out is precisely so that a second operation of the same kind can be performed on that which has climbed out. Thus, when one reaches the conclusion, for example, "all horses are breathing," one finds that this predicate, too, inheres in its subject *per se* and has climbed out. In this way, Grosseteste bases the propositional requirement that premises be *per se* on the metaphysical foundation of inherence. If, however, the metaphysics of perseity is one of inherence, how do we come to know that the predicate inheres in the subject? The epistemological ground of the conclusion is found in the relation that obtains between the premises. Since the premises do not have a similar ground, their ground comes from the fact that the subject inheres in the predicate. But how is such a metaphysical relation grasped?

Grosseteste and *intellectus*

The problem of grasping the principles of a science results from the conditions that these principles must satisfy—the very conditions analyzed above. For if the principles are prior to, truer than, and more known than the conclusion, the disposition by which we grasp these principles must similarly be more certain than that by which we grasp the conclusion of a scientific syllogism. It seems unlikely that we would have such a disposition and yet not know that we have it. However, it also seems unlikely that such a disposition is innate in humans from birth. Aristotle's solution to this problem was to say that the disposition that grasps principles arises out of another disposition that we do possess from birth, namely, sensation:

> It is evident, then, that we are not of such a nature as to possess them [from the start] or to acquire them without the possession of any knowledge or of any habit. So we must have some kind of power, but not such which is more honorable in accuracy. Now this appears to be the case in all animals, for they have an innate discriminating power called "the power of sensation" (*APo* 99b131–36).

Grosseteste understands Aristotle to be looking for a disposition [*habitus*] that can lead to the disposition of grasping principles. Now this prior disposition, Grosseteste argues, must be passive.[14] For if it were active, "it would be more noble and better and more certain than the actual habit of principles" [esset honorabilior et melior et certior quam habitus actualis principiorum . . .] (404).The strategy that Grosseteste maps out here is to show that while the disposition *intellectus* is not innate, it arises from another disposition, which is innate.Thus, *intellectus* arises from sensation.

Sensation is a form of cognition that is of particulars and is "apprehensive of singulars." This "cognitio sensitiva" is a "cognitio apprehensiva," which has as its proper object singular things.The language of "apprehension" will set the standard for discussions of *intellectus* for later thinkers.[15] For Grosseteste, this apprehensive knowledge is a "receptive, sensual potency."[16] While any given individual sense (e.g., sight) is apprehensive, the "sensus communis" is judicative.[17] This pairing of an apprehensive cognition and a judicative cognition both revolve around singulars and both seem to be treated as "receptive" by Grosseteste.

When sense apprehends something, rational creatures are able to retain the sensed forms [*formae sensatae*], to collect from many sensed forms one memory, and from many memories to raise up an experience [" . . . sed in rationalibus iam contingit ex multis memoriis excitata ratione fieri experientiam. . . ."].[18] Grosseteste recaps this movement from sense to experience and adds a final step: "From sense, therefore, a memory is made, from many memories an experience is made, and from the experienced universal, which is outside particulars, not such that it is separate from particulars, but it is the same as them, the principle of arts and sciences arises" [Ex sensu igitur fit memoria, ex memoria multiplicata experimentum, ex experimento universale, quod est praeter particularia, non tamen separatum a particularibus, sed est idem illis, artis, scilicet, et scientiae principium].[19]

Grosseteste's position ends in a similar ambiguity to that in which Aristotle's ends. Just what are these universals that stand outside particulars but that are not separate from them? Earlier in the text, Grosseteste seemed to identify such universals with Platonic forms. However, he also argues that they are causative in an efficient way of the particulars.This allows Grosseteste to make good on Aristotle's claim that the premises contain the cause of the conclusion. His realism—that is, his positing of universals as somehow separate from particulars—leads him into this Aristotelian ambiguity. How are these universals discovered? Here, because of his insistence on the role of the receptive, sensual potency, he seems to say that they are somehow derived from singulars. But without prior knowledge of such a universal, how can some universal be found?

Grosseteste's ontology is clearly Neoplatonic, as we shall see in more detail momentarily. This leads to an emphasis on both the real, independent existence as well as the causative function of universals. Yet does this ontology not require an epistemology of a kind that Aristotle is attempting to overcome in *Posterior Analytics?* Grosseteste's analysis is even more troubling than Aristotle's because of its emphasis on the role of a sensory grasping of particulars. He, more than Aristotle, seems unable to resolve this with his ontology of universals. This will be one of the main areas of contention in the development of the question of the scientific status of theology and the role of *intellectus* therein. Therefore, a brief glance at Grosseteste's analysis of universals in *scientia* would be helpful.

The Ontological Description of a Universal

Grosseteste's first extended treatment of the ontological nature of universals comes in the tenth conclusion he adduces from Aristotle's *Posterior Analytics:* "Every demonstration is from incorruptibles" [X conclusio est hec: omnis demonstratio est ex incorruptibilis. . . .].[20] This question arises because singulars are corruptible and if one is prior to the other and ceases to exist, it is impossible that the other remains. However, universals are discovered in such corruptible singulars. Furthermore, demonstration is possible of such universals that are discovered in corruptible singulars. Thus, Grosseteste must here address the issues of the nature of these universals, their relation to corruptible singulars, and how they come to be known.

Grosseteste makes four attempts to explicate the relationship between universality and incorruptibility. His first attempt is from the perspective of the perfect intellect:

> Universals are principles of coming to know and according to an intellect that is pure and separated from phantasms, they are able to contemplate the first light, which is the first cause, they are the principles of knowing the *rationes* of uncreated things which are existing from eternity in the first cause. Indeed, cognitions of created things which had been in the first cause eternally are the *rationes* of creating things and the formal, exemplary causes and they themselves are also creatrices. And these are what Plato calls ideas and the archetype of the world, and these are, according to him, genera and species and principles of being as well as of coming to know, because they are intuitive, since the pure intellect is able to concentrate on these, the pure intellect is able to know the created things most truly and manifestly in these, and not only created things, but also the first light itself in which it comes to know the others.

> [Ad hoc dicendum quod universalia sunt principia cognoscendi et apud intellectum purum et separatum a phantasmatibus, possibilem contemplari

lucem primam, que est causa prima, sunt principia cognoscendi rationes rerum increate ab eterno existentes in causa prima. Cognitiones enim rerum creandarum que fuerunt in causa prima eternaliter sunt rationes rerum creandarum et cause formales exemplares, et ipse sunt etiam creatrices. Et he sunt quas vocavit Plato ydeas et mundum archetypum, et he sunt secundum ipsum genera et species et principia tam essendi quam cognoscendi, quia, cum intellectus purus potest in his defigere intuitum, in istis verissime et manifestissime cognoscit res creatas, et non solum res creatas, sed ipsam lucem primam in qua cognoscit cetera.][21]

In this description of the nature of universals, according to the pure intellect, we see several interesting features that these universals possess. First, these universals, when contemplated through the first light,[22] are the principles of coming to know the uncreated *rationes* of created things. That is, through these universals, one can come to know the ideas or, as Grosseteste says, "archetypes," that are existing from all eternity. In a sense, these universals are the plan for created things. In this way these universals are not created by God, but rather, exist in God from all eternity. They are the truth of the created things. Therefore, if one wishes to know most truly the created things, one must look to these archetypes. If one begins with the created things themselves, one is only looking at a simulacrum of the ideas. These ideas, then, are prior both temporally and metaphysically to those things that they cause. It is easy to see how causality is tied to universality on this level. This link provides a certain kind of necessity in causation itself. For the divine ideas are both universal and the archetypes of creation.

Grosseteste describes here a relation between the reasons [*rationes*] of things and the things of which they are the reasons. The *rationes,* according to him, are eternal. They are nothing less than the divine ideas, the archetype or the plan that God has for the creation of the universe. These, then, are the causes of the created things. As eternal, they would be necessary— not, though, in the sense that God had necessarily to create *these* things, rather than others. God, presumably, would have ideas for all things, whether created or not. So while they are in themselves necessary, they cannot account, in themselves, for the necessity of the things of which they are ideas. Grosseteste, however, never admits that possibility.

Therefore, when one begins to look for the causes of something being what it is, one will have to look at these divine ideas. The necessity here is twofold. On the one hand there is a necessity that describes the connection of the divine ideas to the things created below them. These ideas exist in God from all eternity. Hence, they are incorruptible. They are in themselves necessary. On the other hand, since the created things find their reason for being in the divine ideas, the divine ideas must find their reason for being only in God. This is precisely the second direction of necessity. In this case, since God is in no way contingent or mutable, the divine ideas have precisely the same necessity that God has. This element of "borrowing being" is typical of Neoplatonic cosmology.[23] The two directions, then, can be stated as "from the top downward," and "from the bottom upward."

The problem, of course, is to reconcile these two directions—a task that Neoplatonism always must solve. But the reconciliation is not an easy one. For we have here an attempt at synthesizing two distinct orders: the order of genesis and the order of knowledge. On the side of the order of genesis, these divine ideas, the archetypes of the created world, are responsible for all the things that the human intellect can discover in the world. Thus, they are the causes of the world. However, they are not, ultimately, seen as the efficient causes of the things, but rather only as the formal causes. The question, though, is whether or not the same causal connection between these divine ideas and the created things can be seen if one begins not with the ideas—because such a beginning is not available to the human intellect—but with the very created things themselves.

These two directions of necessity point to the two directions that can be traced in the cosmology introduced here. For, as we have seen, Grosseteste is here dealing with the pure intellect, that is, the intellect that is able to see these ideas in the first light. Therefore, these intellects are able to see the universals from the side of their origin in the creator, because they are illuminated by the light of the creator. However, for us in this life ("who run short of purity"), there is no ability to intuit immediately the first light.

There is an interesting problem here that is precisely the difficulty raised by the role of *nous* in Aristotle's *episteme* as it is inserted into the context of a creative God: A level is posited to which we have no access, and yet Grosseteste himself is aware of this level, at least in a methodological way. That is, Grosseteste posits a type of knowing (that of the pure intelligences), to which, on his own admission, he himself is not privy. Therefore, there must be some systematic requirement for this level of knowledge. This level allows a cosmology that is contingent from our point of view but, nevertheless, in itself is necessary. The necessity flowing from the divine ideas is met with the contingency of our world as we grasp it. The positing of a level of knowing that we can never ourselves achieve means that the question will always be resolved in terms of necessity. The systematic problem that this level attempts to solve is precisely the same as that solved by *ordo* in Aquinas: If the cosmos is to be grasped rationally at all, then it must exhibit necessity.[24]

We who fall short of purity are able, though, to receive irradiation from the created light, that is, the light that is after the first light, and it is through this inferior light that we are able to know the posterior things.[25] This is the second way of describing the connection between universal and incorruptible "ideas" and the corruptible world that is experienced first in the order of knowing.

Notice, however, that in this attempt to work back upward in the opposite direction a gap is encountered. This gap cannot be filled with rational steps up the causal chain. Rather, this gap must be filled with "divine illumination." The gap experienced here is enormous. For Grosseteste argues not only that divine illumination is required to get a glimpse of the divine ideas that are the causes of the sensible, corruptible world, but *further* because these ideas are the causes of the created world, the human intellect *needs* divine illumination in order to know the things posterior to the uncreated ideas. Thus, the problem is not how to reconcile the order of genesis, which begins with God and the divine ideas and proceeds downward, with the order of knowledge, which begins with the created things and moves upward. Rather, the problem is how to begin, in both cases, with the very divine ideas and work in both directions at the same time. It is precisely this problem that the metaphysics of light attempts to solve in its combination with a theory of irradiation or illumination.

The causal nexus always proceeds downward, each prior level being the formal, exemplary cause of the posterior level. Now the light that comes from the first level, that is, from the primary light, proceeds into the being of the corporeal species. And it is here, at this point, that the ideas—now understood as created ideas—are the principles of knowing according to our intellect, that is, the intellect that is irradiated by the created light. Obviously once the ideas have been located as the focal point of both the order of genesis and the order of knowledge, the problem arises that the order *above* the ideas (namely that pointing first to the intelligences and ultimately to God) and that *below* the ideas (namely that of the created, sensible, corruptible world) are not ontologically compatible. The attempt to link these two levels should hinge on a central point. However, in Grosseteste's analysis, this central point proves unable to provide the link that would make these levels compatible.

It should happen that this central point should share some characteristics with both levels.[26] Here, Grosseteste denies the ontological continuum in favor of a continuum of light. That is to say, he leaves undeveloped the order of genesis and instead develops the order of knowledge. For the human intellect can get to know the divine ideas because we can receive illumination, not from the first, uncreated light, but from a second, more contracted, created light. Thus the ontological continuum is replaced by a continuum of light, which is in the end an epistemological continuum. This light will always lead back to its source, and thus provide access to that level that is the cause of the created, inferior things. This solution to the problem, however, leaves the ontological nature of the level of the ideas unresolved.

The fact that Grosseteste makes this move toward light and, therefore, toward epistemology shows that his fundamental concern is with making the cosmos intelligible. The condition for the possibility of making the cosmos intelligible is seen, by him, to be necessity. This requires, in turn, an ontological ground that would offer some *things* that are, in and of themselves, necessary. However, rather than turning toward ontology, Grosseteste turns directly toward that cosmology. This is where his method and that of Aquinas reveal common concerns. Though each in his own way, both find the need to infuse the cosmos itself with necessity in order to make it intelligible. This means, then, that the very necessity of the created cosmos must rest in God. Yet, this cannot happen completely, for revelation, in some way or another, is required in both thinkers. Revelation, then, serves both methodologically as that through which a contingent cosmos is shown, from the point of view of the creator, to be necessary.

Let us proceed to the other descriptions of these universals before explicating this problem in relation to the theory of science. According to the third description, celestial bodies are the causes of terrestrial, corruptible things. Now the intellect is not able to contemplate the incorporeal light itself. It is, however, able to contemplate the causal connections that obtain between the celestial and the terrestrial spheres. These causal connections are again the principles of knowing and of being of the terrestrial world, and in themselves, these causes are incorruptible.

The fourth way is that a thing is known through its formal cause (141). However, formal cause must be interpreted in a very special way in order for it to account both for necessity and for incorruptibility. Here, Grosseteste must interpret Aristotelian form in precisely the same fashion as he has interpreted light, that is, as the archetype of created things. Thus, form becomes something more like a formal cause of the things, discovered only through the created things themselves, and not directly. Grosseteste seems to think that the analysis of universals from the point of view of the pure intellect is compatible with the analysis from our point of view. In this way, Platonic forms would be the result of the former, while Aristotelian forms would be the result of the latter. However, the former has shown precisely the inadequacy of the latter to reach its desired goal: knowledge of the true nature of universals.

The contradiction here is apparent. For the task to which Grosseteste set himself was to ground necessity in universality.[27] Now, however, it seems that direct access to this necessity cannot come from within the human intellect alone, but only from divine action. Thus, it is only through some activity outside science that science can gain its required necessity. The problem here, however, is that Grosseteste attempted to explicate the universal in such a way that it would match up with his metaphysics of light, which requires illumination. He failed to see that these two directions cannot be maintained at the same time. In this way, we are able to see the difficult role that these universals come to play in his science.

Here we must notice the difficulty with which Grosseteste is struggling: Aristotelian science doesn't work with a Platonic ontology. For the requirements of per-se (i.e., essential) predication are such that the universals must come from within the subject itself, otherwise the construction of a syllogism would do no good. In other words, the ontological gap between the subject and predicate of all propositions is such that there is no room for universals that exist outside the particulars involved: They must be discovered *inside* the universals. Grosseteste has already told us that the truth of a proposition is known when an identity in the substance of the subject and of the predicate is seen, and the necessity of a syllogism is known when the identity of both of the extremes with the middle is seen. Now because of this, it became clear to him that the universals of which Aristotle speaks cannot have existence outside the subjects and predicates of the syllogism. The fact that such a relationship obtains between the subject and the predicate, that is, such that the universal must be discovered within the subject itself, shows that this interval cannot be filled by the very Platonic type of ideas that Grosseteste has argued that universals must be.

It seems, then, that we have two interpretive options. Either, on the one hand, Grosseteste held that there are two types of universals, that is, (1) species and genera; and (2) these archetypes of the created world. These two universals do in fact relate—indeed, a metaphysics of light seems the best way to allow these two types of universals to coexist. But they carry with them different ontological characteristics. The first type have a completely separate existence in themselves, and cannot be discovered directly from the phenomenon that are observed. The second type, however, are found within the very things of the created world. And it is precisely these that can enter into the demonstrative syllogism.

Grosseteste refers to the relation of these universals found within the things to the things in which they are found as *egredior* (to climb out).[28] He has at the same time, however, raised an ontological problem for himself. For *egredior* is such that there is no gap between that which climbs out and that from which it climbs. However, there is a gap between the subject and the predicate terms of the conclusion. And this gap is filled by the middle. Thus, the problem of choosing a middle (described in Bk. II of *Posterior Analytics*) is the problem of discovering precisely that entity that can fill the ontological gap between the two.

The problem for Grosseteste is that his Platonic ontology, even transformed into a metaphysics of light, is an ontology that produces entities that are "too big" to fit into the ontological gap, and too separate to exist in the relation of *egredior*. In order to fit these separate, archetypal entities into the relation of *egredior,* one would have to interpret *egredior* as an epistemological category. In this way, one would see this relation as a relation through which one comes to know these universals. But the fact that one comes to know these universals through the relation of *egredior* says nothing about the way in which these universals exist in themselves or their ontological nature.

Thus, the problem we noticed above, namely that there is a divergence in the order of genesis and the order of knowledge, appears once again in the multiple ontological descriptions that Grosseteste gives to his universals. On the one hand, one could say that this is a direct result of his combination of a sort of Platonic ontology (clothed in a metaphysics of light) with a sort of Aristotelian theory of science. The demands of Aristotelian science require, then, that Grosseteste modify his overly "large" ontology such that it can fit the ontological gap he sees in Aristotle's theory of science. But surely Grosseteste himself was aware of the problem, for the gap is not so small that he couldn't have noticed it.

On the other hand, one could say that the problem is between an overly formalized theory of causation that results from the metaphysics of light (such that Grosseteste can only speak of "formal, exemplary causes") and the formal-efficient causal nexus that is found in the Aristotelian theory of science. The difference being that in Grosseteste's metaphysics of light, causation comes in two related ways: (1) through the formal causation of the divine ideas (admittedly, passed via light); and (2) in the very contraction of that light into created things. But this is a model of causing simulacra (i.e., *imagines* or *vestigia*). The Aristotelian theory, on the other hand, attempts to relate formal and efficient causation. This means that causal *activity* (which is completely absent from a more Platonic-leaning system, even one clothed in a metaphysics of light) is derived from the essence of the cause. Furthermore, when one attempts to discover why a particular subject has a particular attribute, that cause will come from the form (or essence) of the subject itself, and not from its relation to some higher ontological level of ideas.

Finally, one could attempt a reconciliation along the lines of a discovery of forms in the very things that leads to an understanding of the forms that are the very ideas of God. This path, however, is cut off because of Grosseteste's insistence that some illumination is required in order to understand the things posterior to the forms. In other words, the forms cannot be discovered "from below," but rather the things below are discovered only after having been illuminated from above.

The result is that Grosseteste ends up with two worlds, the connection between them being accounted for only outside of Aristotle's scientific theory. One the one hand, one has the world of things posterior to the created light. In this world, an explication of necessity according to what can be called a "normal Aristotelian" approach works quite well. Above the created light, one has the world of the first light coming from the divine ideas. In this world, an Aristotelian account of universals and also of necessity does not work—for the simple reason that this is not an Aristotelian world!

However, by attempting to reconcile these two worlds in his metaphysics of light, Grosseteste creates a fundamental problem. The necessity in the world above is too strong (which relates to the fact that the ontology is "too large") and the necessity in the world below is too weak. But in this attempt at reconciliation, Grosseteste, perhaps unknowingly, pointed out a problem that will concern us as we turn to subsequent thinkers. In order to understand the nature of universals and, through them, the nature of the necessity involved in demonstrative syllogisms, Grosseteste felt his only recourse was to cosmology. In other words, the demonstrative understanding of why a certain fact obtains finds its ultimate resolution in the nature of the entire cosmos. This link between science and cosmology is not seen in Aristotle. Although the problem finds an unhappy solution in Grosseteste, it was perhaps his insight to see this relationship.

Conclusion

Grosseteste's analysis of the role of universals in *scientia* points, on the one hand, to a need to assert that the divine ideas, that is, the archetypes of creation, are accessible to human reason. On the other hand, his inability to reconcile the order of coming-to-be with the order of coming-to-know highlights the irreducible character of the existence of the singular. The struggle implicit in Grosseteste's analysis of *scientia* is that of maintaining the intelligibility of the cosmos in the face of the "corruptible" particulars with which our knowledge begins. His solution is to turn Aristotle's universals into Platonic forms and call these "divine ideas." Thus, *scientia* must include a grasp of the divine plan for creation. These divine ideas become the metaphysical basis of *scientia*.

There are two problematic consequences to this analysis, however. The first is that all knowledge will require a kind of natural theology because all knowledge requires recourse to the divine ideas. The second is that while these divine ideas are the causes of things, they are, even for Grosseteste, only grasped through divine illumination. Such illumination itself exceeds the capacity of reason. The recourse to illumination at the deepest level of causal explanation points to the fact that the singulars themselves, whose account was to be given by recourse to the divine ideas, are excessive. Divine illumination points to the distance that still remains between our rational account of the world in which we find ourselves and the existing singulars that make up that world. Divine illumination appears because the order of divine ideas never meets the order of existing singulars. The fact that illumination exceeds rationality points to the fact that the existing singulars also exceed rationality.

Grosseteste's attempt at a solution would bind God's creative activity to the created universe—thus inserting the kind of necessity that reason demands. This means that the excessive character of existing particulars would also disappear into the rationality of the divine ideas. Divine illumination, then, points to the impossibility of this task.

CHAPTER 3

AQUINAS AND THEOLOGY
AS SUBALTERNATE SCIENCE

Introduction

In the preceding chapter, we saw that while Grosseteste did not address himself to the question of the scientific character of theology, his interpretation of Aristotle through Neoplatonism led him to the position that *all* knowledge is ultimately theological. Aquinas seems not to have had access to Grosseteste's work, and thus does not address himself directly to Grosseteste. However, Aquinas does ultimately end up with a position somewhat close to that of Grosseteste. That is, while Aquinas often speaks of the role of revelation in theology, while he often protests that knowledge of God is impossible, while he often states that predicates are not univocally predicated of God and creatures, he ends up with a great deal of positive, natural theology and an insistence that in some sense all knowledge is ultimately theological. In order to trace this through Aquinas's thought, we need first to explicate his argument that theology is a subalternate science. This position depends on Aquinas's understanding of three important issues: (1) self-evident truths; (2) the *scientia beatorum;* and (3) the *scientia dei*. After these three aspects of Aquinas's argument are delineated, we will then be in a position to see just how much knowledge of God is not only possible, but necessary for him, and the consequences this has for his understanding of the relation between reason and the existing singular.

Per se in Aquinas

In q. 1 art. 2 of the *Summa Theologiae,* Aquinas addresses the question of "Whether Sacred Doctrine is a Science." He argues that it is a science, but that the genus of science is twofold. On the one hand, there are those sciences that proceed from principles known by the natural light of the intellect (such as arithmetic and geometry). On the other hand, there are those sciences that proceed from principles known by the light of a superior science (such as optics, which proceeds from principles known through geometry, and music, which proceeds from principles known through arithmetic).[1] In response to an objection, Aquinas tells us that the principles of any science are known either *per se* or are reduced to knowledge of a superior science.[2] As we shall see, he will ultimately argue that theology is of the latter category, that is, it is a science whose principles are not *per se*,[3] but are known through some superior science. Aquinas does not present an argument here as to why theology cannot be a science that begins with principles that are *per se*. The first article, however, has already prepared us for this move. In that article, Aquinas argued that in addition to the philosophical disciplines, some further doctrine is required. This is because, he explains, "humans are ordered to God as toward a certain end which exceeds the comprehension of reason. . . . This end, however, ought to be known beforehand by humans, who ought to order their actions and intentions to that end" [. . . quia homo ordinatur ad deum sicut ad quendam finem qui comprehensionem rationis excedit. . . . Finem autem oportet esse praecognitum hominibus, qui suas intentiones et actiones debent ordinare in finem].[4] Aquinas shows here that there are some aspects of the divine that simply fall outside the grasp of human reason. These, then, must be revealed by God.

This same theme is picked up again as Aquinas moves toward his proof for the existence of God. For if he is going to prove that God exists, he must show first that God's existence is not known *per se* by us.[5] In order to show this, Aquinas points out that something can be known *per se* in two ways: "in one way, in itself and not for us; in another way, in itself and for us" [Dicendum quod contigit aliquid esse per se notum dupliciter: uno modo, secundum se et non quoad nos; alio modo, secundum se et quod nos].[6] Something is known *per se* when the predicate is included in the definition of the subject. Whenever one knows the quiddity of the subject, one will know such a proposition *per se*. It could happen, then, that one does not know the quiddity of a subject and yet that which is predicated of that subject would be included in the definition of the subject. In this case the proposition would be known *per se* in itself, but not for the one who does not know the quiddity of the subject. This is the case whenever the subject is God. In fact, because any predicate that can be applied to God can only be applied essentially, all true predications whose subject is God are known *per se, but not for us*. As a result, theology, from our point of view, cannot be a science that begins from per-se premises. If it is to be a science, it must begin with premises known through the light of a superior science.

In his commentary on *Posterior Analytics,* Aquinas lists several senses of *per se*[7:]

1. When something attributed to something else pertains to the form of it. Since the definition pertains to the form and essence of the thing, this mode of being self-evident can be summed up as when the predicate is the definition or part of the definition of the subject. For example, when a line is predicated of a triangle. Here, however, we must take note that this is not *per se* because a triangle is *composed* of lines, but, rather, because line belongs to the definition of a triangle.[8]

2. Secondly, something can be predicated *per se* when the material cause is predicated of its proper subject. For example, when nose is predicated of snub-nose, it is *per se* because nose is the proper matter of snub-nosedness. The reason why this is per-se predication is because the existence of the accident (snubbed-nose) depends on the existence of the subject (nose). In this mode of per-se predication, the subject is posited in the definition of the predicate, which is a proper accident of the subject [*Post. Anal.,* I, X, 85].

3. There is a sense of *per se* that is not applicable to predication, but rather to being. This sense applies to some particular in the category of substance. As Aristotle points out in *Categories,* something in the category of substance is never predicated of some other subject.[9] Therefore, we are dealing here not with something that is *predicated* of something else *per se,* but rather with something that has its own existence within itself [10] [Ibid., I, X, 87].

4. Lastly, the fourth mode of *per se* is that in which something is said to inhere in something else *per se.* This refers to the efficient cause. When the efficient cause is predicated of its effect, in most cases that will be *per se.* It is in most cases, however, because we must exclude accidental predications. For example, "While this person is walking, it thunders" is not *per se* because it is not on account of walking that there is thunder. However, "this killed thing died," is *per se* because it is on account of being killed that the subject died [Ibid., I, X, 88].

Obviously the third mode of *per se* is of no use to demonstration because it does not deal with predication, and therefore is of no use in propositions that might form a premise or conclusion of a demonstration. In terms of the conclusion, however, both the second and the fourth modes are required. The conclusion of a demonstration is a proposition in which an attribute [*passio*] is predicated of its proper subject (Ibid., I, X, 89). The subject, however, is not merely posited in the definition of its accidents, but is the cause of them. Therefore, we are involved here in the second and fourth modes of per-se predication.

When it comes to the premises, however, we are on a different footing. Here, it seems, we need the first mode as well. For a demonstrative syllogism will demonstrate that an attribute belongs necessarily to its subject. In that case, we will need to show that an attribute belongs necessarily to some other subject, and that other subject can be predicated *per se* of the first: "All animals breathe and all horses are animals, therefore all horses breathe." Here we see the need for a proposition in the first mode of per-se predication. In predicating animality of horses, we are forming a proposition that expresses the essence of the subject—animality belongs to the essence of horse. This is precisely what makes, according to Aquinas, a proposition *per se* according to the first mode.

We have already seen, however, that Aquinas will not grant that theology has propositions that are known *per se* by us. What Aquinas denied is that we can know propositions that are *per se* according to the first mode when God is the subject. We do not know the "what it is," the "quid est" or quiddity of God and, consequently, cannot know any proposition that is *per se* according to the first mode. Aquinas argues, however, that we can demonstrate the existence of God if we start from another position, namely, God's effects.[11] This type of predication, that is, a predication of an effect of its cause, is a per-se predication according to the fourth mode. What is more, if the existence of God is demonstrable, and if science is nothing more than a demonstrative syllogism, then theology is scientific. Thus Aquinas has already gone beyond what he argued could happen in a science of theology. He has found a way to have a science without appeal to revelation, as he argued must be the case with theology. This move—the move toward natural knowledge of the divine—was already prepared for in the first question, if only we are attentive to his argument.

Theology as Subalternate Science

While Aquinas argues that theology is not a science that has principles that are *per se,* he also argues that it is not necessarily the case that *all* sciences begin with premises that are *per se.* There are some sciences, for example, music and optics, that begin not with self-evident premises, but rather with premises that have been proven by a different science. Music is fundamentally based on principles that come to it from mathematics (ratios that form certain chords and so on). Music, then, is said to be inferior or subalternate to mathematics. Mathematics, in turn, is said to be superior or subalternating to music. Thus, not every science must begin with self-evident premises.

If theology is to be a science, it certainly cannot be a science that begins with self-evident premises, according to Aquinas. On the other hand, what science could be superior to theology? Aquinas's reply seems somewhat odd: Theology is a science subalternate to the science that God and the blessed have.[12] How does this relationship work? "From this it follows that just as music believes the principles handed down to it from arithmetic, in exactly this way sacred doctrine believes principles revealed to it by God" [Unde sicut musica credit principia tradita sibi ab arithmetico, ita doctrina sacra credit principia revelata sibi a deo].[13]

This claim is made through the introductory word, "*unde,*" "whence" or "from which [it follows]." The claim that sacred doctrine functions in the same way as music is drawn directly from the previous claim.[14] Let us look at this previous claim in more detail. The genus "science" is twofold: (1) there are those sciences that proceed from principles that are known [*nota*] through the natural light of the intellect; (2) there are those that proceed from principles known [*nota*] through the light of a superior science. The first class contains those sciences that begin with self-evident principles. The second class contains those sciences that begin with principles that are proven in a superior science.

There is here a double ambiguity. The first comes when Aquinas says that a science can begin with principles that are known [*nota*] by the light of the intellect. This type of knowledge ought to be called "understanding" [*intellectus*] and is a different type of knowing than science, which comes through demonstration. They differ in several ways. Understanding is immediate knowledge of a proposition. Science is knowledge that is not immediate because it is mediated or "middled" in a syllogism. What is known by understanding cannot be demonstrated, nor can that which is demonstrated be known through understanding. The two are, quite simply, radically distinct forms of grasping. The distinction is crucial for Aristotle's theory of science to succeed.

Aquinas produces some confusion, however, by using the term *nota* to encompass both sorts of knowing. This is not, in and of itself, problematic. For although the two types are arrived at in different ways, the result, that is, the *knowledge* that is obtained, will be similar. Yet, for the purposes of showing whether or not theology is a science, it is crucial to draw a distinction between *intellectus* and *scientia*. The requirement that premises be grasped through *intellectus* is dropped for subalternate sciences only because it is fulfilled in the subalternating science. Aquinas's use of *nota* to cover both kinds of knowledge, however, leads to confusing the role that *intellectus* would play in an inferior science. An inferior science is inferior not because there is no *intellectus,* but rather because the principles it uses are demonstrable in another science. The relation, then, between the superior and the inferior is crucial for the notion of inferior sciences. By using one term to cover both *scientia* and *intellectus,* Aquinas, in some sense, covers over the relation that an inferior science has to its superior. If we trace the principles back to the superior science, we will find that they are demonstrated in that science, but they will be demonstrated by way of premises that are grasped through *intellectus.*

Theology, as we have seen, cannot begin with self-evident principles. Therefore, theology is an inferior science. The problem here is that the fact that theology *is* a science needs to be proven before we can then go on to ask *what kind* of science it is.[15] Aquinas here asserts that theology is a science and then proceeds to show that it is a subalternate science. The argument, however, that it is a subalternate science, requires the prior assumption that it is a science, which has not yet been proven.

Secondly, when Aquinas talks about the principles of the superior science from which the inferior science proceeds, he uses the phrase, "*ex principiis* notis *lumine superioris scientiae*"—from principles *known* by the light of a superior science. Yet in the phrase following "*unde*" he states the inferior sciences "believe" [*credere*] the principles handed down to them by a superior science. Believing and knowing, however, are two different cognitive states.[16] Furthermore, an object that is believed cannot also be known at the same time, nor can an object, once known, be believed. Both knowledge and belief are "habits" [*habitus, hexis*], that is, cognitive states that assent to a proposition. Aquinas strictly distinguishes these two modes of assent:

> The intellect assents to something in two ways: in one way it is moved to assent by the object itself, which is known either self-evidently (as is obvious in first principles, of which there is understanding) or through something else (as is obvious with respect to conclusions of demonstrations of which we have science); in the other mode, the intellect assents to something, not because it is sufficiently moved by its proper object, but through a certain voluntary choice inclining it more to one part than to another. And if this inclination would come with some doubt and fear from the other part, there will be opinion; if, however, there would be certitude without any such fear, there will be faith. . . .

> [. . . assentit autem intellectus alicui dupliciter: uno modo, quia ad hoc movetur ab ipso objecto, quod est vel per seipsum cognitum (sicut patet in principiis primis, quorum est intellectus), vel per aliud cognitum (sicut patet de conclusionibus, quarum est scientia); alio modo intellectus assentit alicui, non quia sufficienter moveatur ab objecto proprio, sed per quamdam electionem voluntarie declinans in unam partem magis, quam in aliam. Et si quidem hoc sit cum dubitatione, et formidine alterius partis, erit opinio; si autem sit cum certitudine absque tali formidine, erit fides. . . .][17]

Here faith is understood as that state in which the intellect assents to that which is believed. This is strongly contrasted with both *scientia* and *intellectus,* which are distinguished by the fact that they are moved by their proper objects. In the previous passage Aquinas asserts that an inferior science *believes* the principles handed down to it from the superior science. This, on his own account, is to speak incorrectly. But how could he have done otherwise? He was forced into this position by another linguistic "slight-of-hand."

Recall that the passage has told us that "just as music believes the principles *handed down [tradita]* to it from arithmetic, in exactly this way sacred doctrine believes principles *revealed [revelata]* to it by God." Aquinas needs to place both inferior sciences (music and sacred doctrine) in the state of belief because he wants us to see that there is little, if any, difference between principles that are "handed down" and those that are "revealed." Aquinas must cover over this difference if theology is to be a science. Yet, covering over this difference means placing all inferior sciences in the realm of faith, because all inferior sciences would merely believe the principles that are given them from the superior science. The reason for this is because faith assents only to that which has been revealed by God (*ST* II-II, 1, 1c).[18] Therefore, if the only relationship theology could have to its superior science, that is, the science that God and the blessed have, could be faith, then that very relationship must also be ascribed to all inferior sciences, *if theology is to be a science.*

Revelation is central to theology because of the difference that obtains in theology between propositions that are *per se* and those that are *per se notae* for us. If we compare theological propositions to those of other sciences, this problem stands out clearly. In all other sciences, per-se propositions are not simply found, already there, but are formulated through the process of *intellectus.*[19] This process begins with sensation (usually vision) of a singular. After sensing many such singulars, the intellect abstracts a universal. From this universal, propositions can be formed that are *per se* and necessary because they belong not to this or that particular but precisely to this universal that, because it is abstracted, pertains to all the particulars of that universal. Therefore, per-se propositions are *formed* and because of that formation are also *per se notae.* In theology, however, propositions are simply found because they are revealed by God directly. This opens the possibility, discussed above, that there could be per-se propositions that are not *known* to be such by us. Revelation, therefore, is the name of the situation in which we find ourselves to have discovered per-se propositions (though not exclusively per-se propositions) even though we have no *intellectus* of them.

The result of all of this is that Aquinas has placed the entire function of inferior sciences in jeopardy. They all, then, function based on the fact that they believe the principles handed down to them from the superior science. This certainly breaks the close connection that seems to obtain between, for example, music and arithmetic or optics and geometry. No longer are inferior sciences linked to their superior sciences in a chain of inferences. A moment of hesitation, of uncertainty, is now allowed between an inferior science and its superior. If a special exception is made only for theology, such an exception would be tantamount to the claim that theology is not a science. Aquinas claims that this subalternate science is inferior to the science that God and the blessed have. We must now turn to each of these "sciences" in order to see in what way they can be subalternating sciences to theology.

Creative Science, Receptive Science

God's Science

Why does Aquinas turn to the science of God and the blessed in order to ground theology as an inferior science? Perhaps metaphysics would have been a better choice. What is the peculiar nature of the knowledge that God and the blessed have that is helpful to theology? It should not be thought that the "*scientia dei*" and the "*scientia beatorum*" are the same *scientia*. In fact, one wonders whether this name "science" ought to be applied at all to the knowledge that either God or the blessed have. For, as we have seen, science is the cognitive mode by which knowledge of the conclusion of a demonstrative syllogism is known. Such knowledge is *caused* by the syllogism itself. It hardly seems fitting that such knowledge, that is, knowledge that is caused by a demonstrative syllogism, be applied to God. Yet Aquinas is convinced that God has "science." If this is the case, this science cannot be caused by a demonstrative syllogism because that would mean that there would be something outside of God that could cause God.

The solution to this rests in a dictum that occurs throughout Aquinas's writings in various forms: "*scientia est secundum modum cognoscentis*" (*ST* I, 14, 2 ad.3). Science is not the same for all beings but depends on the "mode" of the knower. It is clear that God and humans would have different ways of knowing, and therefore, science would be different for each. The term "science," however, is not purely equivocal because it points more to a continuum than to a radical difference between our science and God's. For, according to Aquinas, knowledge is directly related to immateriality.[20] This is the case because we can distinguish the knower from the known in terms of the limitation or contraction of the form. The thing known has nothing else but its own form, while the knower has in addition the form of something else: "the species of the known is in the knower" [. . . species cogniti est in cognoscente].[21] Thus, a thing is known more perfectly in inverse proportion to the materiality both of the thing known and the knower. Since God is entirely immaterial and knows things separated from all materiality, God knows in the highest degree.[22]

In terms of the knowledge that God has, *scientia* must not mean "demonstrative syllogism." Certainly God does not know through syllogisms. Such a syllogism presupposes discursive reasoning—that is, going from the knowledge of one thing to knowledge of another. This is not possible for God, who knows all things at the same time (*ST* I, 14, 7). Aquinas himself admits this problem when he treats of God's science in *ST* I, 15, 1. His solution to the problem there is threefold: First he argues (though not explicitly) that a thing is known according to the mode of the knower; second, he argues that whatever perfections exist in creatures, the names of those perfections can be attributed to God, who would possess them in a higher and more perfect degree; third, he argues that whatever exists dividedly in creatures exists simply in God. What is clear, however, is that the name "science" is taken from our science and only applied to God because it is a perfection. In this way, there is no argument that God's science is in any way similar to ours. Indeed, his main point is that because of the difference between God's being (without matter) and ours (mixed with matter), the way in which we have science will necessarily have little in common with the way God has science. The application of the term "science" to the knowledge that God has happens in a completely different way. Science is, for us, the highest and the most certain form of knowledge. Therefore, since whatever is best must be applied to God without any limitation, science must also be applied to God.

The reason why Aquinas wants to apply the term "science" to the knowledge that God has becomes clear in answering the question of "whether the science of God is the cause of things" (*ST,* I, 14, 8). "Natural things are middle between the science of God and our science" [. . . res naturales sunt mediae inter scientiam dei et scientiam nostram].[23] Here we see the whole power of Aquinas's argument. With God's science, on the one hand, causing the things, and our science, on the other, being—at least in some sense—caused by the things, Aquinas is able to establish a direct link between our science and God's science. This link is creation. Creation stands between God's knowledge as cause of things and our knowledge as caused by things.

What is perplexing is that in order to see how theology could be a subalternate science we are referred to the science of God as its subalternating or superior science. The opening of the *Summa* led us to believe that if there were to be any connection between God's science and ours that connection would be *revelation,* for God "exceeds human comprehension." When we learn that theology is a subalternate science, we are now led to uncover just what the "science of God" is that is superior to our science of theology. It is God's creative knowledge that forms an opposite pole to our knowledge of the things. Revelation, then, no longer serves as the connection between the subalternating and subalternate sciences. Rather, it is creation that forms the connection.

In his analysis of "*per se*" we saw that Aquinas had argued that God is not *per se for us* and, therefore, a proposition that had God as its subject could not form a premise of a scientific syllogism. We saw, however, that it is precisely because of this that existence can be proven of God. The possibility of a demonstrative syllogism, that is, the possibility of a *science* of God's existence rested on the notion that God's effects could serve as the middle term of a demonstration.[24] Even though the resulting demonstration is not the highest form of demonstration, it is a demonstration nonetheless.[25] The effects of God form the middle term of a demonstration of God's existence precisely because the effects stand as the middle between our knowledge and God's knowledge. Whereas the initial diagnosis of the problem of the scientific nature of theology pointed us toward revelation, Aquinas's own analysis of the *scientia dei* points toward a natural knowledge of God through the effects of God's creative activity.

Aquinas has pointed us to God's science as one science to which theology is subalternated. We began with the notion that not all truths that are required for human well being are accessible to human reason. There are some truths that "exceed the comprehension of reason." Because of the excessive nature of such truths, revelation was seen to be required. Revelation points to the fact that there is a gap between truths that are *per se* and those that are known to be *per se* by us in this life. Theology begins with per-se truths, but these truths cannot be known *per se* by us. Using revelation as a link, Aquinas can establish a relationship between the science of theology and God's science such that this relationship is analogous to that between a superior and an inferior science (e.g., that between music and arithmetic).

When God's science is investigated, however, it turns out that it is the cause of things. The things are seen to be the middle between our knowledge and God's. Because God's science now has a relation (causality) to the things that are the effects of God, the things can be used in place of direct knowledge of the divine essence in a science of theology.[26] So while Aquinas had initially argued that the grasping of the principles of theology could only come through revelation, he now shows us that in fact God's essence can be grasped through knowledge of things that are caused by

God. For whatever God is, God is *essentially,* and if God is the first efficient cause, God is so essentially.

The result of this long journey is that God's role as cause of the world has become the central factor in Aquinas's account of the science of theology. Our science and God's science are linked through the created things. But this requires that creation itself take on the character of rationality—creation has to be understood as causality. God's creative activity must be brought under the domain of reason if Aquinas's journey is to reach its destination. That is, the will must follow the intellect, as Aquinas argues.[27] God's creative activity would then also presuppose God's intellectual activity. This means that God is related to things as the artisan to artifact.[28] But production of this sort requires a grasp of the intelligible form of things. Therefore, God's science is that which grasps the intelligible forms of all things according to which all things are made.[29] Such intelligible forms are precisely those that ought to be grasped in human knowledge.

God's creative activity, however, is not such that God wills whatever is willed through necessity.[30] As a result, the world is contingent, to a certain extent. Yet once this world is created, there is a relationship between this world and God that allows our reason to trace the reasons of things back to God. For the principle of intelligibility to which the divine will is conjoined, is precisely the same principle of intelligibility according to which we can know the world. This principle of intelligibility is nothing other than the divine essence itself. Thus the world is grounded in the divine science that provides the "reasons" for all things.

Furthermore, the world exhibits an order that results from the divine causality itself. This order is the result of the unicity of the ground of all things.[31] If more than one world is posited, according to Aquinas, then the causality of the multiplicity would be chance—that is, each world would be without reason. Our unique world, however, is the result of divine causality stemming from divine wisdom.[32] *That* a world is created at all is contingent. But once a world is created through the divine intellect, then a necessity is introduced such that the order of that world can be traced back to its reason in God's science.

Aquinas has, therefore, led us on a path from theology as a subalternate science requiring revelation, through the science of God, and ending with a rational connection between God and world that has no need for revelation. The fact that revelation recedes from Aquinas's actual practice should also bear itself out in the relation between reason and existing singulars. We will have to trace this effect, but let us now turn toward the other science to which theology is subalternated, the science of the blessed, to see what its role is.

Blessed Knowledge

While appeal is made to God's *scientia* in order to link our science with God's through creation, appeal is made to the science that the blessed possess for a completely different reason. The structural link that occurs between God's science and ours allows our science to function with necessity and, therefore, with certainty. It links our science with the creative activity of God. The link with the science that the blessed have, however, guarantees the starting point of our science.

There is one and only one thing that separates human beings in the natural state from human beings in a blessed state: seeing the essence of God.[33] It is precisely this knowledge that is entirely lacking in any science humans in the present state could have of the deity. Without a proper definition—which is a formula of an essence—of God, there can be no science. Therefore, the knowledge that the blessed have of God is exactly what our theology needs.[34]

In this way, then, if our theology can be attached to the theology that the blessed possess, we would be able to have a theology that would be properly scientific. In other words, we would be able to formulate demonstrative syllogisms that could use God's essence as a middle term to prove that some attribute belongs to God's essence. Without such knowledge, our theology could never attain to the status of science.

There are two interesting features, however, to this appeal to the blessed. The first is the reliance on vision. It is precisely a vision of God's essence that we expect in beatitude. The grasp of per-se premises in *intellectus* begins with vision. The beatific vision would be the only way for humans, therefore, to have *intellectus* of God. To be sure, there is bodily vision and there is intellectual vision.[35] In this division, however, an interesting feature arises. For Aquinas, "visio" is primarily said of the act of sensual seeing. It is only by extension that it is applied to intellectual vision. Why does the term get extended in this way? It is "on account of the dignity and certitude of this sense."[36] Thus corporeal vision, the primary significatum of the name "vision," is a rather potent and certain way of grasping. This, in turn, is applied to other forms of grasping that do not involve the senses, but that are also potent and certain. Here Aquinas does not distinguish intellectual from sensual seeing in terms of a bodily and an intellectual component to vision. They are distinguished in terms of primary and analogical significate. Corporeal vision here, obviously, has pride of place.

The blessed do not have corporeal vision of God, who is in no way corporeal. The beatific vision is an intellectual grasping of the object, which would be in all ways like corporeal vision except that any material element (either on the part of the object seen or the seer) is lacking.

The beatific vision is the natural end of a rational being.[37] As such it is a purely intellectual act—the act of grasping God's essence. Even though it is an intellectual act, the beatific vision is worthy of the name "vision," because, ostensibly, of its certitude. It is precisely this certitude that Aquinas wants to import to our science of theology. Here we return back to the question of self-evidence. "All science is had through some self-evident principle, and, by consequence, through vision" [. . . omnis scientia habetur per aliqua principia per se nota, et per consequens visa . . .].[38] Aquinas here implies a sequence through which something is known first through vision and then is known to be a self-evident principle.[39] Since, however, we lack any sort of vision (i.e., either intellectual or corporeal) of God in this life, we cannot have a science of theology. This vision, however, is precisely that which the blessed have. Therefore, if vision is the prior condition for knowledge of self-evident truths, then we need the vision that the blessed have in order to grasp the self-evident truths that would form the basis of our science of theology.

Science, Vision, and Self-Evident Propositions

Aquinas's appeal to the science that God and the blessed have conceals, to some extent, his concern with the role and function of *intellectus* in the science of theology. God's knowledge of Godself, that is, the science that God has, is a sort of *intellectus* of deity because God knows primarily only Godself. Everything else that God knows is known through the divine essence. Our knowledge of God, however, is not of this kind, nor could it ever be. What we hope for in the life to come is *visio* of the divine essence. It is vision, as we have seen, that lies at the heart of Aquinas's understanding of the procedure of *intellectus*. The self-evidence of the premises of a science, then, comes through vision. This is precisely what is delivered in the science that the blessed have.

Properly speaking, neither the blessed nor God has any science whatsoever, for neither God nor the blessed reason through discourse. This is what science does—it runs [*discurrere*] from the knowledge of one thing to the knowledge of another.[40] The appeal to the science that the blessed have is an appeal to the *visio* that would be required for *intellectus* of a proposition that would have the divine essence as its subject. The appeal to God's science is an appeal to some way of transferring the *intellectus* that is lacking in us from the divine to us. Revelation would mark the site in which the world that is open to rational comprehension remains without ground, without reason, without why. The fact that Aquinas erases revelation, therefore, points to the fact that our reason is capable of grasping the ground in the divine reason.

Since Aquinas is content with analyzing theology as an inferior science, he sees no need to probe the way in which vision stands at the ground of science. This has significance beyond the mere question of whether theology is a science or not. As we have seen, the term "science" does not fit comfortably on the type of knowledge that God and the blessed have. Since neither God nor the blessed *reason*, that is, go from the knowledge of one thing to the knowledge of another through *discourse*, the science they possess is of an altogether different kind than our science, as we saw above. The question of *intellectus* is, similarly, a different one where the science that God and the blessed have is concerned. For us, the question of *intellectus* takes on a propositional quality that is not involved in the science of God or the blessed. The divine intellect has no need of propositions and discourse because it apprehends in one intuition ["*uno intuitu apprehendens*"] everything through the divine essence.[41]

The reliance on a science that is not discursive as the superior science to theology means that Aquinas does not need to formulate the procedure by which one moves from vision to word to proposition, from thing to essence to formula to proposition. It comes as no surprise, then, that Aquinas does not raise the question about how one goes from knowledge of something in vision to knowledge of a proposition that is necessary, universal, and *per se*.

Aquinas had pinpointed the problem with theology being understood as a science: The principles of the science "exceed the grasp of human reason."[42] Without *intellectus* (Aristotle's *nous*) of the principles of science, there is no science. Aquinas's appeal to the subalternate character of theology, however, only postpones the problem of the grasping of the principles. As we have seen, his argument that theology is subalternate to the science of the blessed and of God is an appeal to nothing other than the grasping of the principles of a would-be science of theology. Indeed, the knowledge that God and the blessed have is nothing other than *intellectus* of God and deity. We will see that this type of grasping of God will be called *notitia intuitiva*—intuitive knowledge—by Duns Scotus. Aquinas's problem is that in this life we simply do not have intuitive knowledge of God and deity such that we could use that knowledge to form syllogisms that would be demonstrative.

Aquinas had several options at his disposal for arguing that theology is a science—none of them very good. He could have argued, for example, that theology is a science that does not have a certain grasp of its principles. In order to make this argument, however, he would have had to destroy all the other sciences. For he could argue that theology is a science, but a special kind in which the requirement that there be a grasp of principles that is certain is loosened. But to argue that theology is a peculiar and unique science is to argue that it is not a science. He could have also argued that theology is a science and no other science has a better grasp of its principles than theology does.[43] This move would destroy the certainty that science was supposed to achieve.

The fact that Aquinas's solution, that is, to make theology subalternate to the science of God and the blessed, was almost immediately and unanimously rejected led to increasing concern for how it is that one grasps the principles of a science and whether such a grasping of the principles of theology is possible in this life.[44] We will see how Duns Scotus and William of Ockham understood this process. Let us now turn to see how Aquinas himself understood how *intellectus* functions in *scientia*.

Whither Comes intellectus?

Köpf sums up the issue that surrounds the role of *intellectus* in science well:

> According to Aristotle each science presupposes certain principles, which are not further demonstrable. His proof consists in that one returns, in the treatment of a state of affairs, to an always ground laying stage. Yet such a return cannot continue into infinity, but hits a limit, a first proposition, which cannot be further grounded. These principles (*archai, principia*) are true, they are the first that can be known, immediate, the cause or the ground for the conclusion—in short, judgments that stand at the beginning of a scientific deduction.
>
> [Nach Aristoteles setzt jede Wissenschaft gewisse Prinzipien voraus, die nicht mehr beweisbar sind. Ihr Aufweis besteht darin, daß man in der Behandlung eines Sachverhalts auf immer grundlegendere Stufen zurückgeht. Ein solcher Rückgang kann sich aber nicht ins Unendliche fortsetzen, sondern stößt an eine Grenze, an erste Sätze, die nicht mehr begründet werden können. Diese Prinzipien (*archai, principia*) sind wahr, sie sind das erste, was gewußt werden kann, unmittelbar einsichtig, die Ursachen order der Grund von Schlußforgerungen—kurz Einsichten, die am Anfange in der wissenschaftlichen Deduktion stehen.][45]

The whole key, then, to Aristotle's theory of the demonstrative syllogism must be found in the grasping of the principles, the true, primary, and certain foundation of the syllogism. In this way, a science is only as certain as is the grasp of its principles. If uncertainty creeps in at the level of grasping principles, then the conclusion reached will be equally uncertain.

There are three important texts in which Aquinas addresses himself to this problem. In looking at these three texts, we will see that the positions of later thinkers, most notably Scotus and Ockham, do not differ in significant ways from that of Aquinas. Furthermore, when we turn to Ockham's discussion of *notitia intuitiva* we will see that it is exactly this problem to which he addresses himself.

In Libros *Peri Hermeneias* Expositio

Aquinas begins his exposition of Aristotle's *Peri Hermeneias* in much the same way he begins his exposition of *Posterior Analytics,* that is, by assuming that the *Organon* is a unified text and then attempting to argue how each book of the *Organon* fits into that unity.

> According to the Philosopher in Book III of De Anima, there are two operations of the intellect: a certain one which is called understanding of indivisibles [*indivisibilium intelligentia*], through which the intellect apprehends the essence of some particular thing in itself, the other is the intellectual operation of composing and dividing.
>
> [Sicut Philosophus dicit in III De anima, duplex est operatio intellectus: una quidem, que dicitur indiuisibilium intelligencia, per quam scilicet intellectus apprehendit essenciam uniuscuiusque rei in se ipsa; alia est autem operatio intellectus scilicet componentis et diuidentis.][46]

This distinction must be understood as a distinction between the way in which we grasp terms and their referents and the way in which we grasp propositions and their referents. Here, Aquinas says that the grasping of simples (i.e., of terms and their referents) is a grasping of their essences. If this is true, this is no simple operation. For the grasping of essences involves (1) sensation, (2) memory, (3) abstraction. We will turn to this process in more detail below. For now, the important thing to take note of is that there is an operation of the intellect whose function is to grasp simple things and the terms that signify them. On the other hand, then, we have an operation of the intellect whose function it is to grasp the composition and division of these simples[47]—facts (states of affairs) and the propositions that signify them. To these two operations, Aquinas adds a third, "ratiocination" [*ratiocinandus*]. By this operation, reason moves from what is known to what is unknown.[48]

These three operations, however, are not just three operations of the intellect that stand beside one another. These operations are ordered. The grasping of simples is ordered to the grasping of complexes "because there is not able to be composition and division without apprehension of simples" [Harum autem operationum prima ordinatur ad secundam, quia non potest esse compositio et diuisio, nisi simplicium apprehensorum].[49] There must be some grasp of the terms of the proposition and the things to which the terms refer if there is to be some proposition that is made up of these terms.

In a similar way, then, the second operation, that is, that of grasping facts and propositions, is ordered toward the third, that is, moving from the knowledge of one thing to the knowledge of another. What is needed for there to be ratiocination is the knowledge of a true proposition.[50] Thus, to move from the knowledge of one thing to the knowledge of the other there are two requirements: (1) knowledge of simple things; (2) knowledge of an affirmative or negative proposition to which the intellect assents. From this, the intellect can proceed to "accepting with certainty something which was unknown."[51] Furthermore, Aquinas argues that once the intellect forms a proposition that is based on this grasping of simples, the intellect will assent to that proposition. While Aquinas does not specify how that would happen, we are led to believe that it has something to do with the way in which the simples are grasped or is a direct result of this grasping of essences.

In Libros *Posteriorum Analyticorum* Expositio

When Aquinas comes to the same problem in *Posterior Analytics,* that is, how the *Organon* forms a unity and what role the *Posterior Analytics* plays in that unity, he returns to exactly this relation. The division here is a little more complex, owing to the fact that it is meant to explain the role of *Posterior Analytics* rather than *Peri Hermeneias.* For within the three acts of the intellect mentioned in the *Peri Hermeneias,* another division appears that pertains to this third act, the act of ratiocination. The first two acts appear just as they had in the previous text.[52] However, the apprehensive act is here described not as being of essences, but of what the thing is (*quid est res*). This can be seen, however, as roughly equivalent in that the quiddity of a thing is the same as the essence of the thing. We will, however, return to this interpretation in a moment.

The second operation of the intellect, composing and dividing, too, appears as it did in *Peri Hermeneias.* In *Posterior Analytics,* though, he stresses the fact that in this act, truth and falsity appear where they did not in the first act. Still, though, these two acts are contrasted with the third act, which now takes on the name "judicative."[53] This means that we are led to call the first two acts *together* "apprehensive" while the last act is called "judicative."

Aquinas compares the three acts of reason into which the third act— ratiocination—is divided into acts of nature. The division is based on whether or not there is a "defect in the truth" [Est enim aliquis rationis processus necessitatem inducens, in quo non est possible esse ueritatis defectum. . . .].[54] The defect in the truth can only come about, however, because there was truth there originally and, in the process of ratiocination, that truth can become defective. Where does this original truth come from? It must come from that act of the intellect in which truth is grasped, that is, the second apprehensive act of composition and division. By reserving the term "judicative" for only the third act of the intellect, that is, of ratiocination, Aquinas has left himself with treating the second act of the intellect, that is, of composition and division, as an "apprehensive" act. Indeed, the grasping of simples as well as of propositions are called operations of the *intellect,* whereas the act of judging is called an act of *reason.*

When we look, then, at these two passages in combination, we can see how Aquinas envisions their relationship. First, there is a grasping of simples [*incomplexi*]. This grasping has to be of such a sort, however, that when a simple thing (e.g., Socrates, whiteness) is grasped in this way, we can immediately form a proposition (e.g., 'Socrates is white') and know whether this proposition is true or false. Thus, the apprehension of a complex (e.g., Socrates is white) is dependent upon the apprehension of simples (Socrates, whiteness). Not only must it be dependent, however, but the apprehension of simples must allow the intellect to know the truth or falsity of the ensuing proposition ('Socrates is white'). Thus, the grasping of a proposition is an apprehensive act and not an act of judgment. Only when I have apprehended terms such that I assent to a proposition that affirms or denies can I then form syllogisms about them. The rules of these syllogisms, then, would help to preserve the truth of the original apprehension.

Principles and Science

What remains, then, is to see how this last relationship operates—how one goes from the apprehension of a proposition that is based on the apprehension of its terms to the conclusion of a demonstrative syllogism. The problem here is that the premises of a demonstrative syllogism must be universal and necessary, and yet the apprehension that Aquinas discusses in the *prooemium* deals with particulars and contingents. How does one go from contingent propositions about particulars to necessary propositions about universals? This is the problem of *nous* or *intellectus.*

In explicating the origin and function of *intellectus,* Aquinas follows Aristotle's text quite closely.[55] It belongs to all animals to have a co-natural potency for judging things that are able to be sensed. This potency is called "sense." This is a potency that cannot be acquired or developed, but one that follows upon the very nature of being an animal. Now in certain animals, some impression of the sensible thing remains when the sensible thing itself is absent. Animals in which this is not possible can only have cognition of a thing while it is being sensed. Now the holding of these impressions in the absence of the sensed object happens outside of sense itself. These impressions, that is, the cognition of the thing in its absence, takes place in the soul.

Among these animals that have memory, certain of them are able to apply reason to the impression that remains in the absence of their objects. What this application entails is a drawing together of many memories into one "experience" [experimentum]. This experience must be something that unifies these many impressions into one whole—that is, that draws these many impressions into many impressions of the same thing. In an experience, one impression is drawn together with another—and this is an activity that is proper to the faculty of reason. Aquinas gives an example that will become the standard example used in discussions of the procedure of intellectus: Someone records (remembers) that such an herb has often cured many people of fever and draws from these many impressions the experience that herbs of such a species simply cure fever. However, there is not yet the principle of science, only a contingent proposition that deals with particulars. Reason[56] does not deal with particulars, even with many particulars. Rather, it is the task of reason to draw out of many particulars one common thing. This process belongs, then, primarily to the expert in a particular science.[57]

What is it that the expert of a science does? The expert can draw from these multiple apprehensions of singulars and propositions about them (or states of affairs made up of singulars) some universal proposition that is "without consideration of some singulars" [. . . et considerat illud [i.e., unum commune] absque consideratione alicuius singularium . . .].[58] From a proposition like the one given above—'Such an herb is curative of fever'—the expert can formulate the proposition 'All herbs of this species cure fever.'

As we have seen above, Aquinas returns to this relation in the Summa Theologiae. While discussing the object of faith, Aquinas asks whether faith is "vision." There, he distinguishes between two sources that lead the intellect to assent. On the one hand, there are those in which the object moves the intellect to assent, either through itself or through another. On the other hand, there is that habit in which the intellect assents by voluntary choice. In the first division falls intellectus, which deals with the principles of science, and scientia, which deals with the conclusion of science. The first two, intellectus and scientia are equated with vision. The second, fides, is not related to vision.[59]

Here what is interesting is the introduction of the concept of voluntary choice or act of the will. Aquinas seems to assume that the assent that goes with the grasping of first principles is such that the there is no activity of the will involved. That is, the intellect, immediately upon grasping the principle, assents to it. This can only result from the fact that its truth is guaranteed. But what is the guarantee of its truth? It can be nothing other than the grasping of the essence or the "*quid est res*" of the thing—that is, the apprehension of simples. Here, however, the "*quid est res*" cannot be understood as essence alone. For if this type of apprehension can lead to the proposition 'Socrates is white' then this type of apprehension need not be of the essence of Socrates. In this case, "what the thing is" can mean, "that the thing (Socrates) is white." Indeed, to grasp the essence of Socrates would have nothing to do with grasping the truth of the proposition 'Socrates is white'.

Aquinas equates this grasping with vision: "[E]very science is had by some self-evident principle, and, by consequence, vision. And therefore whatever things are known, there is in some way vision."[60] The knowledge, then, with which science begins is ultimately traceable to vision. Here the term "vision" neatly gathers together two things that had been separated. On the one hand, it was standard for theologians to talk of the beatific vision, based on the biblical idea that there are those who will "See the face of God and live." On the other hand, there is the philosophical preference for vision in theories of truth—whether Aristotelian or Augustinian.[61] This means that the type of vision that forms the center of *intellectus* (later to be called by Scotus *notitia intuitiva*) will begin to take on the characteristics of the beatific vision. The qualities that pertain to the beatific are precisely what Aquinas wants as the basis of science: a grasp of something such that (1) the intellect knows "*quid est res*," which must include properties, qualities, *and* essence and (2) there is more in the grasp than what can be put into a proposition—the object exceeds any proposition that is formed about it. If I apprehend Socrates, there are, perhaps, infinitely many propositions that I would then be able to form based on that apprehension: 'Socrates is white', 'Socrates is snub-nosed', 'Socrates is bald', etc. The initial grasp of Socrates, however, does not specify any of them in particular, yet it must include all of them. If the apprehension of the simple 'Socrates' did not include all of them, how else would I apprehend the proposition 'Socrates is white' and know that it is true?

The issues, as Aquinas has framed them, revolve around my assent to a proposition. Ultimately, for Aquinas, the cause of that assent is vision that apprehends simple things in such a way that when it then goes on to apprehend some complex (a proposition or a state of affairs composed of those simples) it assents to it. Now if such propositions are to form the certain basis of science, then the apprehension and assent to the proposition that is based on the apprehension of simples must be unable to err.

Aquinas implicitly raises two issues that will become central to the concerns over *intellectus*. First, he shows that nothing is in the intellect unless it is first in the senses. This is obvious because no universal is arrived at in the soul unless from some experience that is a collection of impressions that remain from an act of sensing. Second, he shows that while sensation is of particulars, reason is of universals. Yet what he fails to raise is the issue of which acts of sense and, therefore, what kind of experience, allows one to make the crucial move from particulars to universals. Aquinas gives two examples of this move. The first example is when an accident that is universal is drawn from an experience of many particulars that have that accident: From my experience of Plato, not as Plato but as white, and my experience of Socrates, not as Socrates but as white, I am able to focus on the whiteness that belongs to each such that I am indifferent to anything that exists. This indifference to particulars is the beginning of the move to universality. But the result of the indifference is not an essence but an accident that belongs to many particulars.

The second example pertains to the essence of particulars. For if I sense Callias, not only insofar as he is Callias, but also insofar as he is this human, and if I sense Socrates in the same way, that is, as this human, my intellective soul is able to consider the "human" in each and thus arrive at the universal "humanity" or "human." This requires, however, that I focus, according to Aquinas, on the "universal nature" of each. For if I apprehend only that which belongs to the particularity of each, in no way would I apprehend the universal nature in the particular, and, consequently, no cognition of a universal would be caused by my sensing.

In citing these two examples, Aquinas has presupposed what he set out to show, namely, that we can move from our sensation of particulars to positing some universal. Through this last example, Aquinas wants to show that there is an act of sensation in which one can move from particulars to universals and an act of sensation in which one cannot leave the realm of particulars. How does one know, however, whether the requisite act of sensation is present? One knows, according to Aquinas, if one is able to reach the "universal nature" that is in the particulars. One must already have a grasp of what these universal natures are before one can be sure whether an act of sensation can lead one to such universal natures. The test of whether one has *intellectus* and, therefore, certitude, is whether or not one achieves a "universal nature" that must be known in advance.

When the question of necessity and certitude is raised by Aquinas, it is solved also in a peculiarly circular fashion. Since *intellectus* is posited by Aristotle as the only habit that is more certain than science, it follows that *intellectus* must pertain to that which is always true and can never be false. Therefore, whenever one has *intellectus* of some truth, then that truth is known with certitude. How does one know whether one has *intellectus?* It can only be when one has achieved the proper "universal nature" of the thing sensed.[62] Aquinas seems to have failed, then, in his endeavor to show how vision grounds the certitude of science.[63] For it is not vision proper, but rather the universal natures, posited in advance of vision, that ground the certitude of science.

This, however, is not to attack Aquinas by saying that he was not a nominalist. For the argument is not that there cannot be any universal that exists independently of singulars. Rather, Aquinas's own claim was that such universals are found only in particulars. Yet, he has found it necessary to posit such universals *prior* to their discovery in particulars so that one may know when the proper *intellectus* is achieved. Certainly one could hold that the universals are discovered only in singulars without positing their prior existence. This circular chain of grounding, however, will continue to plague medieval attempts to ground the certitude of science in vision, as we shall see in subsequent chapters.

Because of Aquinas's emphasis on universal natures or essences in the process of *intellectus,* the question of vision of particulars is resolved into the ontology of universal natures. Yet, as we have seen, this does not answer the question of how one moves from the vision of a particular to the discovery of a universal nature. Aquinas does not interest himself in the formation of propositions out of terms that would refer to things. Rather than ground per-se propositions in some logical fashion, he seems to think that the universal natures, once discovered, are the ground of per-se propositions.

Once later thinkers, however, begin to see that the problems involved in Aquinas's argument concern the subalternate character of theology as a science, they will be forced to rethink the question of *intellectus.* The way in which vision of sensible particulars can begin a process that will culminate in universal, necessary, per-se propositions will become the central issue in the question of whether theology is a science. The ingredients of those discussions, however, are already found in Aquinas.

Aquinas and the Negative and Analogical Character of Theology

The conclusions drawn here concerning Aquinas's philosophical position have, to a large extent, ignored one feature of Aquinian thought that has, since his death, become the focus of considerable attention, namely, the role of negative theology and analogy in Aquinas. Indeed, the conclusions drawn here seem to fly directly in the face of Aquinas's own statements regarding what can be said of God. As a result, before turning to medieval responses to Aquinas, some word must be said about the role of both the *via negativa* and analogy.[64]

Immediately after proving the existence of God, Aquinas begins the next question, on the divine simplicity, by stating, "Having known of something whether it is, it remains to inquire how it is, such that it would be known of it what it is. But because we are not able to know of God what God is, but what God is not, we are not able to consider how God is, but more how God is not." [Cognito de aliquo, an sit, inquirendum restat, quomodo sit, ut sciatur de eo, quid sit. Sed quia de Deo scire non possumus, quid sit, sed quid non sit: non possumus considerare de Deo, quomodo sit, sed potius quomodo non sit.][65] Aquinas then goes on to say that the next thing that must be considered is how God is not. The first move in this direction is to remove any composition from God. This will yield the seemingly positive result that God is entirely simple.

The first way of removing composition from God comes by negating that God is bodily. For if God were or had a body, then God would be a composite that would include that body and some other principle in addition to that body. For this reason, in *ST* I, 3, 1c he engages in three negative arguments showing that God is not corporeal. First he argues that no body moves if it is not itself moved. Since God has been shown to be the immobile mover, it is clear that God does not have a body.[66] Second he argues that since God is the first being [*ens*], potency can in no way belong to God. Since every body is in potency, God cannot have or be a body.[67] Lastly, he argues that a living body is more noble than a nonliving body, therefore that through which a body lives is more noble than the body. Since God is the most noble of all, God cannot have or be a body.[68]

Clearly, each of these arguments proceeds negatively. It begins with some characterization of what belongs to a body, qua body, and it then goes on to show that the conclusions already reached concerning God lead immediately to the denial of that characteristic in relation to God. Each argument also relies on a proof for the existence of God.[69] These arguments show quite clearly what Aquinas has in mind when he speaks of knowing how God is not. Yet when Aquinas goes on to argue that God is perfect, this negative method of demonstration is no longer used, but Aquinas now turns to a positive proof.

The difference between "body," for example, and "perfection," for example, is the difference between a name that is said of God negatively, that is, by removing something from God and a name that is said of God absolutely and affirmatively. Aquinas, unlike Maimonides, never denies that we can predicate names absolutely and affirmatively of God: "Certain names of this kind signify the divine substance, and are predicated of God substantially, but fall short of the representation of God" (*ST* I, 13, 2c).

These names, however, signify according to the mode of our understanding. The problem is whether names that signify according to our mode of understanding, which falls short of a proper understanding of God, can signify God—and signify substantially:

> God has preeminently in Godself all perfections of creatures as simply and universally perfect. From this it follows that any creature to the extent that it represents God, and is similar to God, to that extent has some perfection, not nevertheless such that it represents God as something of the same species or genus, but rather as an excelling principle, from whose form the effect falls short, yet whose effects follow some kind of similitude: as the forms of inferior bodies represent the power of the sun. (*ST* I, 13, 2c)

The reference to preeminence and to the relation of an effect to its cause is central to Aquinas's argument that names predicated of God and creatures are predicated analogously. For according to Aquinas such names are predicated of God and creatures only because creatures are ordered back toward God "as toward the principle and cause in which all perfections of things preexist in an exceeding way" [. . . ut ad principium, et causam, in qua praeexistunt excellenter omnes rerum perfectiones].[70] Therefore, even though we do not have positive knowledge of God's essence, and although the names predicated of God come from creatures, we still are in the midst of a positive theology because those names already arise out of the fact that creatures are referred back to God as to a preeminent cause and principle.

Now it is precisely this relationship between cause (even if preeminent) and effect that has been at issue in the reading of Aquinas presented above. For the causal relation between God and creatures must stand outside of the scope of Aquinas's negative and analogical theology because these presuppose it. For the notion of causality is already included in the proofs for the existence of God, and Aquinas's negative and analogical theology rests on those proofs. It is in unpacking this notion of causality in relation to creatures that Aquinas comes to posit divine ideas as exemplary causes of singular things. Those arguments, in turn, are more difficult to read analogically, and impossible to read negatively.

Finally, it is well known that Aquinas's theory of analogy was opposed by both Scotus and Ockham, and most who follow in their tradition. As a result, both Scotus and Ockham refuse to read Aquinas's arguments on the basis of analogical predication. Looking back at Aquinas from the point of view of later thinkers, one can see the entire discussion of divine names as a strategy that allows Aquinas to have it both ways. On the one hand, he can investigate all sorts of divine attributes in a positive manner. On the other hand, the moment at which he would be accused of propounding a natural theology or a doctrine of creation as necessary emanation, he could refer back to the merely analogical use of terms.

Conclusion: Reason, Ground, Existence

In the following chapters, we will turn to the problem of *intellectus* as it is raised by Duns Scotus and William of Ockham. In order to see Aquinas's contribution to those discussions, it is necessary to anticipate some of those conclusions. First, we will see that Scotus will initially use this term "vision" to describe the initial stages of how one grasps the principles of science. In his later writings, he will come to replace the term "vision" with the term "*notitia intuitiva*" in the context of the beatific vision as well as the context of the initial stages of our grasping the first principles of science (*intellectus, nous*). Secondly, we will see that Ockham places his discussion of *notitia intuitiva* in precisely the same context as does Aquinas, that is, in the context of operations of the intellect that science presupposes and that are ordered toward science. The conclusions Ockham reaches will, for the most part, be exactly the same ones we have seen Aquinas reach here. Furthermore, one of the main examples Ockham uses in his discussion of *notitia intuitiva* will be that of a herb curing a disease—the same example that we have seen Aquinas using. Ockham and his readers must have understood that he was attempting to analyze the requirements of *nous* that forms the basis of science.

Aristotle's search for the causal grounds of existing singulars led him to the problematic of the relation between vision, universal essences, and universal truths. From the perspective of *Posterior Analytics,* knowledge is only knowledge of the rational ground of an existing particular.[71] Yet the rational ground is always posited in advance of that of which it is the ground. Because of this, reason cannot be the ground *of* the particulars.

Aquinas did not give up the search for the rational ground of existing singulars. For him, God ultimately provides the ground for existing singulars both in being and in reason. This search ultimately leads Aquinas to hold the untenable: The rational ground of existing singulars (God) must also be grounded in reason. Aquinas seems to threaten the very creative will of the divine.[72] While this is in itself merely a theological question, it signifies a deeper philosophical issue. God's creative will can be read also as a sign of the very givenness of existing singulars. When Aquinas grounds this activity in reason itself, he too gives up the givenness of the phenomena.

Aquinas seems to force us onto one or the other horn of a dilemma: Either we give up the givenness of the phenomena or we give up the search for the rational ground of existing singulars. This dilemma became the condition upon which subsequent thinkers would address the question of whether theology is a science.

CHAPTER 4

DUNS SCOTUS AND INTUITIVE KNOWLEDGE

The Condemnations of 1277
and the Existing Singular

As mentioned earlier, a commission of theologians "and other wise men" drew up, at the request of the bishop of Paris, Etienne Tempier, a list of propositions that could not be held or taught at Paris. Oxford soon followed in handing down similar condemnations. The condemnations focus on propositions of metaphysics, psychology, epistemology, and natural philosophy that seem to arise out of the Latin appropriation of the Islamic, and particularly Averroistic, reception of Aristotle. The condemned propositions concern everything from the agent intellect to the eternity of the world. A good number of the propositions, especially those concerning cosmology and natural philosophy, bring forth condemnation principally because of their limitations on divine power.[1] There are some such propositions that are taken from the work of Aquinas. One such proposition, which would have deep and long-lasting consequences on medieval philosophy, was "God could not make more than one world."[2] This proposition goes to the very heart of the relation between existing singular and rational ground.[3]

The proposition that God could not have made more than one world was asserted by Aquinas on the basis of the intelligibility of the cosmos and on the basis of the intelligibility of God as creator of the cosmos.[4] If more than one world is posited, Aquinas argues, then there is not a cause of the world but only chance.[5] For Aquinas, the unicity of the world works in two directions at the same time. First, it is what allows us to posit the unicity of God as its rational cause. Second, it is what allows us to posit the rational order of the world as being caused by God. In short, the order of the world, for Aquinas, points to there being only one God and the uniqueness of God points to there being only one world.

Aquinas saw that in order for God to provide a rational ground for existing singulars, the creative activity of God had to be given some ground in reason. For if we admit of a multiplicity of worlds, then this world in it-self tells us nothing about the creator other than that the creator created. Without being able to trace existing singulars to a rational ground in the divine intellect, the world appears as chance. Cause is the mode of being of the rational ground, chance is the mode of being without ground. Aquinas, therefore, argues that this is the only world that God has created, and consequently the divine intellect can serve as the rational ground of existing singulars within the world. The rationality of the cause demands that there not be a plurality of effects. The rationality of the cause, in turn, is seen in the order of the world.

The Condemnations of 1277 ushered in the requirement that one pay attention to God's creative will, rather than God's knowing and creative intellect.[6] This has serious implications for the theory of science and for the question of whether theology is a science. It meant that one could no longer posit God as the rational ground of existing singulars, at least not in the way that had been done by Aquinas. From this point on, the ground of the existing singulars had to be found independently of the divine intellect. We turn now to see how Duns Scotus attempts to reformulate the concept of science in such a way that it does not bind God's creative will at all.

Science in Scotus

Scotus's theory of science is remarkable in that it seems to leave the existing singular aside altogether.[7] This move away from knowledge of the singular as existing as the basis of *scientia* is possible only because of Scotus's understanding of the role of the "subject" of a science and the truths that are "virtually contained" within it. Scotus's terminology shifts from speaking of the "subject" of a science to speaking of the "object" of a science, though at times he uses both interchangeably.[8] However, these two terms have different senses that point out different aspects of the role it fulfills. The term "subject" finds its philosophical site in Aristotle's logical and metaphysical writings as that which is able to take predicates. The term "object," on the other hand, finds its philosophical site in relation to a potency (such as sight and its proper object) or as the relation of cause and effect. Scotus likens the "object" of a science to the latter.[9] The object of a science is related to the *habitus* of science as a cause is related to its effect. Each science would have *one* determinate thing that serves both as the subject to which all propositions must refer and as the object that functions as a cause of the habit. These no longer need to be held apart.

Science as such, according to Scotus, is not ordered on a determinate common genus, under which all objects of this science fall, but fundamentally on a determinate unique essence, and this essence, as the first subject, if it is completely known in its virtual implications, would fall together with the first object, which causes the entire habit under consideration and terminates the inclination of a potency.[10]

What is also innovative in Scotus's theory of science is the fact that it almost completely gives up the Aristotelian idea of explanation.[11] For Aristotle, as well as for Grosseteste and Aquinas, science was supposed to explain the reason why a given phenomenon was the way it was. This explanation would come through the causes—which in one way or another would make their way into the premises of a scientific demonstration. For Scotus, the subject of the science is more crucial than the causes of a certain phenomenon. The reason for this has to do with Scotus's conception of what the subject of a science is and how it functions in a science.

For Scotus, science grants certain knowledge of something. This certainty does not come through knowledge of causes, as it did for Aquinas and Grosseteste, but rather through knowledge of the one essence from which all other things treated in a science are ordered. Scotus defines science as "certain cognition of a demonstrated necessary truth mediated by necessary truths which have been previously demonstrated, which is of a nature to have evidence from a prior, evident, necessary truth applied to it in a discursive syllogism" [. . . cognitio certa veri demonstrati necessarii mediati ex necessariis prioribus demonstrati, quod natum est habere evidentiam ex necessario prius evidente, applicato ad ipsum per discursum syllogisticum].[12] The question that this definition seems to avoid is what kind of cognition is required of the "prior, evident, necessary truths," from which the truth under consideration is generated. As we have seen, this is the question of how we get knowledge of per-se truths from which we can demonstrate some conclusion.

Scotus, in fact, worries very little about the question of how we can acquire knowledge of the principles of a science. In his definition of science, the question is one of "evidence." The conclusion of a scientific syllogism is one that "goes begging elsewhere" for its evidence, that is, it is not "self-evident." This ought to lead to the conclusion that the premises of a scientific syllogism are "self-evident." Yet Scotus does not provide much information about how they are self-evident and how we can acquire knowledge of them precisely as self-evident.

Sensation, Intellection, and the Principles of Science

In his *Quaestiones super Libros Metaphysicorum Aristoteles,* Scotus addresses the question of how knowledge of the principles of demonstration are acquired in at least two places. The first discussion comes in the context of the question whether from experience an art is generated. There Scotus argues that the first operation of the intellect is the apprehension of singulars.[13] These singulars can be either sensible or not. If they are sensible, then they are not able to be understood in a universal unless they have first been sensed in a singular. On the other hand, if they are not sensible, still some recourse to sense must be made.[14] Once some singular has been grasped in sense, the intellect is able to engage in composition and division. From this, the intellect is able to arrive at a complex concept, which is known to be true by the natural light of the intellect. Thus we are able to achieve such truths (and these can serve as the "first principles" of a demonstration) from frequent sensitive cognition and memory and experience, "through which we know that the terms of such principles in their singulars are conjoined in reality, just as sense frequently sees this totality and this majority to be conjoined" [Possunt etiam cognosci quia veri sunt ex frequenti cognitione sensitiva et memorativa et experimentali, per quas cognoscimus terminos talis principii in suis singularibus in re esse coniunctos, sicut sensus frequenter vidit hanc totalitatem et hanc maioritatem coniungi].[15] In this way, the principles of a science are grasped through the apprehension of simples and the cognition of truth in composition.

Scotus here introduces a two-stage procedure in the grasping of the per-se principles of a demonstration.[16] In the first stage, a particular is sensed frequently. This first stage includes sensation, memory, and experience, just as in *Posterior Analytics*. At this stage, however, there is no contribution from the intellect. Therefore, all of these aspects (sensation, memory, and experience) have singulars as their objects. The premises of a demonstration, however, are propositions, not singular things. Scotus needs a way to move from these aspects dealing with particulars to a proposition.

Scotus, much like Aquinas, argues that the move from sensation of particulars to propositions comes in a second act of the intellect. In this act, the intellect composes or divides what has been given in the first act. Here, as in Aquinas, truth enters because truth pertains to a proposition, not to a thing.[17] While this act relies to a certain extent on sense, it is also adds something that sensation cannot: assent. This assent cannot be automatically elicited from sense and is a necessary component of the propositions that can be the principles of a demonstration. On this level, the singulars are gathered together not according to some accident, but according to their common nature.

The question is how to move from apprehension of singulars in sense to apprehension of propositions concerning these singulars in the intellect. The common nature, which Scotus raises here, is the rational ground of the singulars. But how do I move from the singulars to the rational ground? What is the act that can take account of the singular while moving at the same time to a more universal level, the level of the common nature? Scotus argues that from sense, the intellect is able to apprehend simples and immediately the most universal, that is, "being" and "thing." That is, when a singular thing is sensed, from this sensation alone, the intellect immediately forms the concept "being" or "thing." When I sense something, I can at least maintain that it is a being or a thing. When simples have been apprehended by sense, either truly or falsely, propositions are formed by the proper power of the intellect. This happens first concerning what is more universal and then concerning what is less universal. From the most universal, common concepts are formed and the intellect immediately assents to them. This assent, however, does not happen on account of sense, because the assent is more certain than is possible from sense. From what is given from sense, the intellect accepts the cognition of the truths of those propositions. The intellect can also form less universal propositions that are also immediate, but are not known immediately or are not known to be immediate because the terms are not known.[18]

Scotus here sets up a relation between two forms of cognition: sensitive and intellective. While the sensitive is responsible for grasping singulars, the intellective can grasp singulars, and, in addition, is able to engage in judgment. Judgment is what allows the intellect to move from the level of sensation of particulars to the level of cognition of immediate propositions. This means that intellective cognition is going to be more certain than sensitive cognition.[19] Intellective cognition is more certain for three reasons. First, intellective cognition can judge how things stand with sensitive cognition, but sensitive cognition does not have that capacity either with respect to itself or with intellective cognition. Second, certitude does not arise in apprehending what is true, unless somehow the truth that has been apprehended is known. Only the intellect can engage in this kind of reflective activity of judging the truth of what has been apprehended. Third, sense never perceives the immutability of the object, even though it may sense that which is immutable. For sensation perceives the object only when it is disposed to sensing that object. Thus sense cannot provide certainty about immutability because it itself will be subject to mutation along with the object.[20]

The lesser certainty of sense derives precisely from its being tied to the object that it is sensing. In this grasp, sensitive cognition grasps a true thing, but is unable to engage in the "doubling" move—that is, judging the thing in relation to something else—that is required in order to grasp "its truth." Yet the certitude that arises when the intellect does engage in this doubling move is a certitude that is not tied merely or only to what is given to the intellect in sensitive cognition. The fact that judgment arises in intellective cognition points to the fact that the certainty and "truth" of this kind of cognition do not belong to the grasp of the existing singular as existing and singular. Scotus here clearly points to the "common nature" as the ground of the "truth of the thing." Yet, sensation is always responsible for giving *something* to the intellect for its activity. The main task is to give sensation a kind of security so that insecurity will not affect the certitude of intellective cognition. This security is given in what Scotus calls "*cognitio intuitiva.*"

Sensitive Cognition and Intuitive Cognition

In question 3 of book two of his *Quaestiones super Libros Metaphysicorum Aristotelis,* Scotus returns again to the question of sensitive cognition. Here, however, he introduces a distinction: In *sense,* one cognition is intuitive, immediately proper; another is immediate and is *per se* proper through species, but not intuitive.[21] For Scotus, the basic distinction is not whether cognition is proper to its object, but whether that cognition comes immediately from its object or immediately from its species. Both modes are per-se of the object. Intuitive cognition is "of a present thing not only through species, nor only under its determination as knowable, but in its proper nature" [Primus est intuitivae cognitionis, quae est de re praesente non tantum per speciem, nec tantum sub ratione cognoscibilis, sed in propria natura].[22] This is contrasted with knowledge of the thing known through a proper species produced from itself. What is the warrant for accepting direct knowledge of the nature of a thing apart from its representing species?[23]

In many contexts, Scotus gives a *theological* argument and a *philosophical* argument for the possibility of intuitive cognition.[24] In the *Quaestiones super Libros Metaphysicorum,* the theological argument is raised only because it arises out of the philosophical. The philosophical argument is based on the notion that whatever perfection is in an inferior capacity, that same perfection must be present in the superior capacity, otherwise the superior would not be superior. Sensitive cognition is able to know something insofar as it is present through its essence. Therefore this capacity must also belong to the intellect.[25] Scotus never denies, as we have seen above, that the senses are able to grasp the thing as it is present. But what do the senses grasp? Is there a grasp of the essence of the thing sensed? Is there a grasp of "its truth" such as comes only through composition and division? In short, how can one move from the sensation of a present thing to the cognition of it in the intellect such that the intellect can grasp the immediate presence of the thing? The intellect is just that capacity for retaining species whether the object is present or not. Therefore, it does not know intuitively, but only through species.

Scotus does not have a strictly *philosophical* answer to this question. This is where he raises his *theological* argument. This type of cognition is precisely what we expect in the beatific vision. If this cognition is possible there, then it is, in general, possible. If it is not possible in this life, then that must be a result of the specific conditions of our way of knowing here and now. But once it has been asserted that intuitive cognition is possible in the beatific vision, the principle follows that for any discrete act of the senses there can be a concomitant act of the intellect *about the same object,* and this act of the intellect can be called vision.[26] We must always keep in mind, however, that we are dealing with a capacity of the *intellect.* Scotus often offers an analogy: Intuitive cognition is to the act of sensing as abstractive cognition is to knowledge through phantasm.[27]

The issue, as Scotus frames it in the *Quaestiones super Libros Metaphysicorum,* is how one moves from sensation to the principles or premises of a science. There are two competing Aristotelian claims within which the issue of the relation of sensation and intellection operate. On the one hand is the claim that nothing is in the intellect unless it was first in sense. On the other hand there is the cluster of claims that knowledge is of what is universal and sensation is of what is singular. What, precisely, is the bridge between sensation of singulars and intellection of universals?

Aristotle, as we saw, held that the process of *intellectus* begins in sensation of the singular, but ends in a kind of knowledge of the universal. These can then form the basis of *scientia.* This science, however, is the attempt to uncover the rational ground, the why, of the existing singular. In this context, we can ask how we can move from a sensitive intuitive act that has an existing and present singular as its object to an intellective act that has a universal, which is the reason for the singular, as its object. Scotus, as we shall see, allows for abstractive knowledge to form the basis of science. If abstractive knowledge abstracts from the presence and existence of the singular that is known in this way, and if abstractive knowledge can form the basis of science, then science ceases to function as the rational ground of the existing singular. Abstractive knowledge is precisely "abstractive" because it is "indifferent" to the presence and existence of its object. The propositions, and science based on them, consequently, will also be indifferent to the existence and presence of the singular. Scotus moves science away from its role of rational ground of what is conceived of the singular. In short, we could formulate a distinction between the *conceptual* ground of our knowledge and the *rational* ground of the existing singular.[28] The existing singular remains still fundamentally outside of the rational ground that science provides.[29]

Abstractive Knowledge of Singulars

When Scotus comes to ask whether the viator can have knowledge of truths that are knowable per-se of God under the concept of deity, he argues affirmatively and deploys the distinction between intuitive and abstractive knowledge to prove his point. "Cognition is twofold: one kind is through a species, which is not of the thing in its presence, and this cognition of the thing is called abstractive; the other is cognition of the thing as it has being in actual existence, and this cognition is called intuitive" [. . . est sciendum quod duplex est cognitio; quaedam quidem est per speciem, quae est rei non in se praesentis, et haec vocatur cognitio rei abstractiva; alia est cognitio rei ut habet esse in actuali existentia, et haec dicitur cognitio intuitiva].[30] Scotus goes on to say that this abstractive cognition applies to knowledge of a visible thing through a phantasm, that is, through sensory imagination. The phantasm does not represent the thing in its presence and actual existence. Indeed, although Scotus does not say so, it is clear that this is precisely the job that the phantasm was to fulfill—its task was to *abstract* from the presence and existence of a singular *in order to* come upon the universal. If phantasms were not already abstractive, then they could not fulfill their task in science.

The issue for science, therefore, is not that it *must* begin with knowledge of some existing singular in its existence and presence, but rather that it must begin with some *distinct* knowledge. If abstractive knowledge can deliver distinct knowledge, then *intellective* intuitive cognition is not a requirement for science. As Stephen Dumont has pointed out, this is precisely the "controversial aspect" of Scotus's doctrine. For Scotus is arguing nothing less than that complete and explicit apprehension of all the necessary features of a nature does not need to come through intuitive knowledge.[31] This means that science does not need knowledge of a thing in its presence and existence, but can be grounded on knowledge that abstracts from its presence and existence. Scotus seems to indicate here, in addition, that such abstractive knowledge does not require intuitive knowledge.[32]

What happens, then, to the singularity of the existing thing? Notice first that what is missing is not the *particularity* of the existing thing. For particularity means to be already an instance of a universal. This knowledge would belong precisely to abstractive knowledge because abstractive knowledge could grasp the existing thing through its common nature. The singularity, on the other hand, is the existing thing as it stands in its existence without its being subsumed under a general concept, without its being an instance of a universal. The singularity, in short, is precisely what is given up in the search for the rational ground.

What is this singularity that escapes abstractive knowledge? For Scotus, individuals are always individuals of some nature. This nature deserves the name "common" because in itself it is neither universal nor singular, but is indifferent to either.[33] If a singular has a common nature that is common to it and other singulars, then what accounts for the unity of the singular such that it is a singular and not the common nature? This unity must be something positive that is intrinsic to the singular. Whatever this is, will be responsible for the individuation of the thing.[34] Scotus rehearses many candidates for that which individuates a thing (e.g., accidents, matter, existence, negation) and finds none of them suitable to the task. His solution is to say that there must be some "formality" in the individual that "contracts" the common nature to individuality. This formality will come to be called "*haecceitas*" by the tradition.

In the *Quaestiones super Libros Metaphysicorum*, Scotus argues that due to the fact that the individual is an individual because of its haecceity, it is not knowable *per se.*[35] What moves our cognitive powers is not the singularity of the object, but the nature of the object. This nature, as we have seen, is indifferent to singularity and universality. Even though it is really the same as the individual, it does not formally move the cognitive potencies *as singular.*[36] The issue here is not about intellects in general, but specifically about our intellects, which are receptive only through some mediating natural action. Angels, for example, can understand singulars because their intellect is immediately receptive of the action of the object. Our intellects require the mediation of sense, which will not grasp the singular *per se.*[37]

Ultimately, Scotus may have changed his mind on the issues of intuitive knowledge of singulars. Certainly Ockham read him as maintaining that such knowledge of singulars is possible—for Ockham it is even necessary. For such knowledge would be required for existential certitude of any fact.[38] Furthermore, in the same question, Scotus shows concern that without such intellective intuition, one would not be able to form propositions and syllogize about contingent things and facts. This is, in fact, a logical consequence of his original insight that science is not about explanation, but about certitude. In that case, we can have *scientia* of contingent truths. This *scientia,* however, must have a grasp of the contingent truth such that its certitude is not open to question. Scotus seems to have realized that only *intellective,* intuitive knowledge would provide the certitude for contingent science. Yet that science itself, on Scotus's own account, is not about explanation, but about unpacking that which is virtually contained in the subject of the science. Even when Scotus brings *scientia* to contingent truths, that *scientia* remains posterior to the singular, contingent thing. This view of science never gets behind the singular to its rational ground, but rather the singular provides the ground for science.

Scotus offers us two sides of the knowledge of individuality. On the one side, the individual, as this highest grade of actuality, must be knowable. On the other hand, what is not knowable about the individual is precisely its individuality. All individuals, therefore, are to some extent only singulars and not particulars. This is precisely why abstractive knowledge of individuals should form the basis of science. For sensory intuitive knowledge can grasp the singularity of the thing. But this singularity is precisely what cannot form the basis of science. Universality is achieved through abstractive knowledge alone. This universality forms the basis of science. This leads to the consequence that for Scotus, science is no longer the search for the rational ground of existing singulars. This can best be exhibited if we return to Scotus's theory of *scientia*.

Types of Scientia

We stated earlier that the main focus of the Scotistic theory of *scientia* is the "primary subject" of the science. Scotus begins his investigation into the concept of the primary subject of a science by distinguishing between two main types of science: in itself and in relation to an intellect of a certain kind. While Scotus in fact only distinguishes between a science in itself and "for us," it is clear that our intellect is but one kind of intellect that could be engaging in science. Therefore, science "for us" stands in for science as it is possible in an intellect of a certain nature, rather than only for humans. Certainly angels and the blessed have science, though each of a particular kind that might differ from science in itself.

Science in itself, then, is the ideal of a science that is drawn from its object: "Science in itself is that which is of such a nature as to be had of its object in the way in which the object is able to manifest itself to a proportionate intellect" [De primo dico quod quaelibet scientia in se est illa quae nata est haberi de obiecto eius secundum quod obiectum natum est manifestare se intellectui proportionato . . .].[39] This presents an "idealized" notion of science in that it describes science in abstraction from all the concrete conditions of an intellect engaging in science. Science "in itself" derives merely from the object of the science and pays no attention to the capacities of any particular intellect to grasp adequately that object. Science itself ought not to be defined only in relation to the intellect and its capacities. When we abstract from the concrete conditions of a particular intellect, or kind of intellect, we are left merely with the conditions that are supplied by the object of the science.

Because we have now understood the distinction between a science and its concrete instantiation in a particular intellect (or kind of intellect), we see the fundamental status of the object or subject of the science. It is this alone, and not the power of the intellect, which determines the science. This primary object is that which "primarily, virtually contains in itself all truths of that habit" [De secundo dico quod ratio primi obiecti est continere in se primo virtualiter omnes veritates illius habitus].[40] That is, the object is such that it "does not depend upon the other things which it contains, but the other things depend upon it" [. . . illud est primum quod non dependet ab alio sed alia ab ipso; ita igitur 'primo continere' est non dependere ab aliis in continendo sed alia ab ipso . . .].[41] In short, science is the intelligible species of its subject.[42]

The work that the primary object will perform rests on two concepts used to explicate the notion of a primary object that Scotus does not here define: the "primariness" of the object and the fact that it contains "virtually" all truths of that science. As we saw above, "primary," for Scotus, means that the object does not depend upon the things that it contains, but those things depend on the primary object. This is what Scotus calls elsewhere an "essential order."[43] In his Parisian commentary, the relation of the truths of a science are related to the primary subject:

> [B]ecause in what is ordered essentially, it is necessary that it is reduced to something which is simply primary; now knowables of whatever science have an essential order among themselves in knowability, because conclusions are known from principles, principles, then, if they are immediate, are known from terms, as has been said, but terms of the principles themselves are known from the concept [ratio] of the subject.

> [. . . quia essentialiter ordinatis necesse est ea reduci ad aliquod primum simpliciter; cognoscibilia autem cuiuscumque scientiae habent ordinem essentialem inter se in cognoscibilitate, quia conclusiones cognoscuntur ex principiis (patet ex dictis) principia tandem si sunt immediata, cognoscuntur ex terminis, sicut dictum est; sed et terminus ipsius principii cognoscitur ex ratione subiecti.][44]

The characteristic of being "first" belongs to the first object only because there is an essential order of all the truths of a science, and that order must point toward some "first." It is in reference to this ultimate first, that is, to the first object of a science, that Scotus grounds all other truths that pertain to the science whose first object it is. The first object provides the conceptual ground for all those truths that are ordered to it.

The notion of the first object as that which makes a science an essential order of truths fundamentally alters the entire notion of science itself. Scotus has traced the ground of science in a reverse order: The conclusions are known from the principles, the principles are known from the terms, and the terms are known from the ratio of the first object. But what has become of the grasp of an existing particular? In this idealized notion of science, the question of the existential status of the first object, and therefore all subsequent "objects," becomes completely irrelevant. All that is relevant is unpacking the conceptual content of the *ratio* of the first object.

For Scotus, an order is essential if it arises out of the essential account or nature [*per se ratio*] of the terms ordered.[45] This *per se ratio* is that "which expresses the essence or quiddity of something without determining whether it has being in the mind or outside the mind."[46] Since science is an essential order that is such only because there is some first, it also proceeds from an essential account or *per se ratio* of that first. Science, therefore, does not necessarily need to begin with a grasp of an existing singular, because it can function as *scientia* without paying attention to the existential status of that first object.

Science, Rational Ground, and Existing Singular

The question of the rational ground of existing singulars, which arises in the question of the scientific status of theology, appears in Scotus in a unique light. For him, science is indeed the uncovering of the rational ground of the propositions that are knowable by science. However, this rational ground does not purport to be the rational ground of existing singulars, but rather it is the rational ground of knowledge itself. The concept of the existing singular forms the conceptual ground of *scientia*.

The innovation of Scotus's position on the scientific character of theology, therefore, rests on the role that *abstractive* knowledge plays in science, rather than on the role *intuitive* knowledge plays.[47] Intuitive knowledge as knowledge of an existing singular in its presence and existence is problematic neither as a way of analyzing the beatific vision nor of the traditional way of understanding *intellectus* as the beginning of *scientia*. This distinction in *ways* of knowing does not presuppose distinct *objects* known. Rather, the distinction, for Scotus, revolves around whether or not a thing is known in its presence and existence.

From the point of view of science, this distinction is only significant in terms of the concept of "evidence." As we have seen, Scotus argues that science has four conditions: (1) it is certain cognition, without any deception or doubt; (2) it is concerning something known necessarily; (3) it is caused by a cause evident to the intellect; and (4) it is applied to the known through a syllogism or syllogistic discourse. The central feature of these conditions is the third. For it is with this third condition, the condition of evidence, that we can understand the first and the fourth. The certainty of the thing known will derive from the certainty of the cause of the knowledge. The cause will come through a syllogism or through *syllogistic dis-*

course.[48] Thus the issue of these two ways of knowing is a question of being an "evident cause" or "evidently causing" knowledge of a conclusion.

Put in this way, the question that Scotus must answer is whether and how abstractive knowledge, which does not grasp the thing in its presence and existence, can provide evidence for the knowledge of a conclusion. The issue of grasping the premises of a scientific demonstration is precisely the issue of securing the ground of the conclusion. The premises must, therefore, be truer and more known than the conclusion itself. Can we grasp premises in this way through knowledge that does not grasp the existing singular in its presence and existence?

Scotus's definition of science and the role played in science by the primary subject do not require the knowledge of the primary subject in its full presence and existence. For Scotus, all the truths of the science, from its nonprimary truths to its ultimate conclusions are virtually contained in the first subject. The science is a *habitus* that is the "intelligible species of the first object" [Ille habitus qui dicitur scientia est species intelligibilis primi obiecti . . .].[49] Ultimately, the knowledge that is gained in science need not be grounded in the grasping of the existence of a singular, but can be grounded in the grasping of an essential nature abstracted from all existence and presence. This is the force of Scotus's distinction between intuitive and abstractive knowledge.

The fact that Scotus is able to base his notion of *scientia* on abstractive knowledge, the role that the primary subject plays in that science, and his insistence on the singularity of the singular move the entire operation of *scientia* out of its relation to existing singulars and into the realm of the *conceptual content* of our knowledge. Science is not left insecure on this model. But the existing singular is left out of its sphere of operation. Science has become conceptualized such that it is nothing more than the logical unfolding of the conceptual content of some primary subject. The logical unfolding follows an essential order, but an order that leaves out the existing singular.

In this way, the singular *as singular* receives a certain prominence. The fact that our concepts of singulars and the conceptual unfolding of those concepts in science leaves the singularity unthematized (or, at the very least, posterior to the singular itself) means that this very singularity is never subject to the universalizing tendency of science. If science remains as the search for the rational ground of the object, it is the search for the rational ground as conceptualized and not as it is in its existence. For Scotus, our concepts are never adequate to the objects as long as we recognize that the singularity of the object resists that conceptualization.

This amounts to a reorientation of science. This reorientation will be pursued by Ockham and subsequent thinkers to its full extent. Where

Ockham—and "nominalism" in general—will differ from Scotus is on the question of the real basis of our general concepts in things. Ockham gives up the existence of "common natures" outside the soul, and thus gives up Scotus's solution to the question of individuation. For Ockham, individuals are simply given. General concepts, therefore, arise out of these individuals without a real basis in those individuals. This will bring the Scotistic primacy of the singular to its logical conclusion.

CHAPTER 5

OCKHAM AND THE NATURE OF SCIENCE

Introduction

As we have seen, an overarching concern of the later Middle Ages in the question of whether theology is a science is a concern with the rationality of the world and how God fits into both that world and that rationality. If science (in any age, in any of its various meanings that share a family resemblance) is supposed to tell us something about the world, then the world itself will feel the changes that the concept of science undergoes. The question of whether theology is a science was central in keeping alive the discussions of Aristotle's somewhat strange concept of science. It is in this question that theology and epistemology, ontology and cosmology fold into one another—as we have already seen. The three questions that have to be asked at the intersection of these fields are: (1) What is God's relation to the cosmos? (2) How is God knowable? (3) How is the cosmos knowable? These three questions are, in the end, one and the same: Is God a part of the cosmos or not? If God is part of the cosmos, then God ought to be knowable by human beings insofar as the cosmos is knowable. There is no natural barrier to knowledge of God when God is part of the world. Often, a supernatural barrier has to be erected. The supernatural barrier then takes the form of a mere prohibition. While God ought to be knowable, it is the case that God is not. Conversely, if God is not a part of the world, then God is, by nature, supremely unknowable.

Ockham certainly introduces a new sort of epistemology, philosophy of nature and metaphysics. Yet it is difficult, if not fruitless, to attempt to place a priority on any of these. Is Ockham concerned with safeguarding God from philosophy? Or is he concerned with the nature of universals? Or is he concerned with a theory of science that would take experiment and experience seriously? Or is he concerned with God's absolute power? All of these questions can certainly be answered in the affirmative. Yet one cannot establish a priority in which one concern is the origin and the driving force of the others.[1]

Such procedures can be avoided. There can be no doubt that whatever the origin, the concerns and consequences of Ockham's thinking are clear: (1) there is a deep concern with the absolute power of God; (2) there is a logical concern with the status of universals; (3) there is an epistemological concern with guaranteeing knowledge while still grounding that knowledge in singulars.[2] The mixture of these concerns produces several far-reaching consequences: theology becomes divorced from *scientia; scientia* becomes divorced from necessary laws; physics becomes divorced from metaphysics. In all of these, Ockham can be seen as unpacking the conclusions already achieved by Scotus.

Ockham on Theology as a Science

Ockham's definition of science is deceptively simple:

> [S]cience . . . is evident knowledge of a necessary truth, which, by its very nature, is caused by premises applied to it in a discursive syllogism.

> [[S]cientia . . . est notitia evidens veri necessarii, nata causari per praemissas applicatas ad ipsum per discursum syllogisticum.][3]

This definition applies to the conclusion of a demonstrative syllogism and not to the syllogism as a whole. We have science of a proposition that can be demonstrated. Ockham delineates three conditions that this definition sets forth:

1. Science is *evident knowledge.* This condition rules out other ways of grasping the conclusion—for example, opinion, suspicion, faith, and so on.
2. Science is of *necessary truths.* This rules out science of contingent truths, which are not knowable by science, properly called.
3. Science is caused by premises. This rules out the possibility that the conclusion can be grasped in the same way in which the premises are grasped.[4] Here, the phrase "by its very nature, is caused" is meant to allow that the knowledge of the same proposition can be caused in other ways, but it is of such a nature that it *can* be caused by premises. It may, for instance, be caused by experience, but it is in principle possible that it could also be caused by premises. There are, presumably, truths that cannot be caused by any premises: for example, 'Socrates is sitting'.[5]

On the face of it, this definition does not seem that much different than Aristotle's or Grosseteste's or Aquinas's. There are, however, features of this definition that are striking. Perhaps the most remarkable of these is the claim that something that is knowable by science is also knowable through experience.[6] There is no doubt that Aristotle, and indeed many commentators on *Posterior Analytics,* argued that the conclusion of a demonstrative syllogism could be grasped without that demonstration. Ockham's example, "the moon is eclipseable," can certainly be grasped by experiencing an eclipse. Ockham gives no indication that knowledge gained through experience would only be knowledge of the fact and not of its reasons.

What is striking about Ockham's definition is that knowledge can be gained *either* through a demonstration *or* by experience and Ockham shows a certain hesitation in ranking these two modes of knowing. That type of knowledge gained either through experience or through demonstration is called *notitia evidens*—evident knowledge. The fact that one proposition can be known in two completely different ways was a matter of some controversy.[7] Ockham follows Scotus in reorienting science away from necessity and toward certainty. By arguing that the same truth can be known either through experience or through science, Ockham places science in a different site and gives it a different role. Ockham asserts here that the same *knowledge* can be caused either by demonstration or by experience. The issue for him, as for Scotus, is *evidence.* Knowledge of a proposition can be made evident either by a demonstration or by experience.[8] Ockham even asserts that there are some truths, in which an attribute is predicated of a subject, that are not knowable by science at all, but only by experience. These would include per-se propositions, but are by no means limited to per-se propositions.[9] This means that for Ockham the most important factor in science is the cause of the *knowledge,* not the cause of the *fact.* Uncovering the reasons, that is, the metaphysical ground, of the object of experience ceases to be the goal of science. Only when science has given up the search for the ground of existing singulars can knowledge gained through experience be put on a par with (and even privileged over) science.

The definition Ockham gives of *scientia* is closely related to the conditions he gives for propositions that are knowable by *scientia:*

> I say that a proposition which is knowable by science, in its proper sense, is a proposition which is necessary, dubitable, able to be made evident by necessary, evident propositions which are applied to it in a discursive syllogism.

> [Circa primum dico quod propositio scibilis scientia proprie dicta est propositio necessaria, dubitabilis, nata fieri evidens per propositiones necessarias evidentes, per discursum syllogisticum applicatas ad ipsam.][10]

Here a different characteristic is introduced: doubt. This will come to be one of the most important features of Ockham's theory of science. That a proposition be dubitable means that it is not *per se nota*.[11] With the definition of science and the conditions of a proposition that is knowable by science in its proper sense, we can take a preliminary look at Ockham's conclusions concerning whether and how theological propositions are scientific.

Which Theological Propositions are Knowable by Science?

Ockham begins this discussion by laying down some distinctions. First, he calls our attention to the difference between a demonstration *propter quid* and a demonstration *quia*. This is a standard distinction in Aristotelian science and it is the distinction between a syllogism that proves a fact (*quia*) and a demonstration that proves the reason why the fact is the way it is (*propter quid*).[12] A demonstration *propter quid* was thought by all to be the highest form of demonstration. Secondly, Ockham states that, "Of those which are predicated of God, certain are things and certain are only concepts" [. . . illorum quae praedicantur de deo quaedam sunt res, et quaedam tantum conceptus].[13] Lastly, Ockham declares that, "With respect to those which are predicable of God, either the thing itself which God is or some concept predicable of that thing is able to be a subject" [. . . quod repsectu praedicabilium de Deo potest subici vel res ipsa quae Deus est vel aliquis conceptus praedicabilis de re illa].[14] Ockham does not state here how a concept is distinguished from a thing, but supposes for the present that the concept is not really or formally the same as the thing of which it is a concept. With these distinctions in hand, Ockham lays out several possible kinds of propositions and analyzes whether they can be known by science.

The first category of proposition is that in which something intrinsic to God is predicated of the divine essence such that the essence in itself would be the subject and something that is really the divine essence would be the predicate.[15] Such propositions cannot be demonstrated. The main reason Ockham gives for this is that such propositions would have to be *per se nota,* and, therefore, not demonstrable. There is no distinction whatsoever between God and the divine essence. Therefore, what is really God is the same as the divine essence. Propositions of identity are the highest form of propositions that are *per se nota.*[16] Furthermore, such propositions are not self-evident for us in this life.

In the second category of propositions are those same kinds of propositions as in the previous case, but which would predicate something not of the divine essence but of the divine persons. These are indemonstrable for the same reason, that is, such propositions are *per se* because of the real identity of the persons of the trinity with whatever can be predicated of it. Anyone who knows a proposition in which one of the divine persons is the subject and something that is really God is the predicate cannot doubt that proposition.[17]

In the third category are those propositions in which a concept that is common to God and to creatures is predicated of the divine essence. Such propositions cannot be demonstrated a priori—that is, *propter quid*. The reason is that such propositions are "immediate."[18] "Every common concept is immediately predicated of whatever is contained in it and into which it is immediately divided" [. . . quod conceptus communes praedicabiles in quid de deo et creaturis non possunt de divina essentia in se demonstrari a priori. Hoc patet, quia quaelibet talis est immediata, quia omne commune immediate praedicatur de quolibet contento in quod primo dividitur].[19]

Animal, for example, is a common concept, that is, a concept that is predicable of many which are not of the same species. Animal divides immediately into rational and irrational. To say that animal divides immediately into these two is to say (1) there is no genus or species between the two divisions and the concept common to them; and (2) the division completely exhausts the common concept (there are no animals besides those that are irrational and those that are rational). When these are predicated essentially (*in quid*) of the divine essence, they are immediate propositions. This holds true for all the transcendentals: wise, good, etc. These divide immediately into created and uncreated. Therefore, such propositions are not demonstrable, but are self-evident—again, not for us, but for the intellect that has intuitive knowledge of the divine essence.

In the fourth category are those propositions in which a concept that is both connotative and negative and common to both God and creatures is predicated of the divine essence. According to Ockham, such propositions are demonstrable.[20] These concepts are predicable of something more universal than God and creatures and this more universal concept can serve as a middle term to demonstrate that such a predicate belongs not only to creatures but also to God.[21] Ockham gives the following example of such a demonstration: 'Every being is good; God is a being; therefore, God is good'. This syllogism, however, must be precisely understood. For when the terms "being" and "good" are taken as being predicated *in quid* of the divine essence, then the syllogism is not demonstrative. Therefore, the syllogism is valid only when "good" and "being" are taken as connotative terms and not predicated *in quid*.

A connotative term is one "which signifies something primarily and something else secondarily" [Nomen autem connotativum est illud quod significat aliquid primario et aliquid secundario].[22] The sign of such terms is that they have a nominal definition only. "White" is such a connotative term. Its nominal definition is "something informed by whiteness" or "something having whiteness."[23] Notice that what a connotative name signifies primarily is not, perhaps, what one would expect. "White" primarily signifies a thing, that is, a subject. Only secondarily does "white" signify the whiteness that this subject has. The term primarily signifies the *something* that is informed by whiteness.

In the case of the demonstration under consideration, we can say that when "good" is taken as a connotative term, it signifies a being, that is, a subject, primarily, and only secondarily signifies the goodness that that being has. The proposition 'Every being is good' is true, according to Ockham's definition of truth, because both the subject, "being," and the predicate, "good," stand for the same thing. This is because the predicate, as a connotative term, must be analyzed into the proposition 'a being that has goodness'. The connotative term signifies "being" primarily and the result is that both subject and predicate stand for the same thing or things.

Every a priori demonstration requires that the predicate that is to be demonstrated belongs first to something other than the subject. This concept must be the middle term that will be used to demonstrate that the predicate belongs to the subject. This condition is satisfied in the demonstration because "good" belongs primarily to being. In this way, "being" can be a proper middle term in the demonstration. Yet, one wonders just what is achieved in such a demonstration. We will return to this point below.

In the fifth category, Ockham considers those propositions in which a concept that is connotative and negative and yet proper to God is demonstrable of the divine essence. Such propositions, for him, are not demonstrable a priori.[24] Examples of such propositions are 'God is creative', 'God is omnipotent', 'God is eternal', and so on. The reason why such demonstrations are impossible can be drawn from what was said above. Every demonstration requires that there be something higher than (i.e., more common than) both the subject and the predicate that can serve as a middle term. In such propositions, there is no such term that can be the middle term.

The difference between this group of propositions and the previous group is that here we are dealing with terms that are predicated of the divine essence and are *proprii*. A *proprium* is an attribute that belongs to one and only one species or genus. The ability to laugh is a *proprium* of humanity. This means two things: (1) *all* humans are able to laugh; and (2) *only* humans are able to laugh. The difference between the fourth and the fifth groups is that when a common concept is predicated, there is something prior to both God and creatures (e.g., "being") that can serve as a middle term. When we are dealing with a *proprium,* no such prior thing is available.

The last group of propositions that Ockham considers are those in which a composite concept is predicable of God essentially, yet it is dubitable whether that concept properly belongs to the divine essence. Such propositions are demonstrable because one could use the divine essence itself, or a distinct cognition of deity, or some common concept as the middle term of such a demonstration. Examples of such concepts are "to be true," "to be one," "to be good," and so on. That is, all the transcendentals are such concepts that can be demonstrated of the divine essence. This is demonstrable because the proposition is able to be doubted. This doubt, however, arises only in an intellect that does not have "intuitive knowledge" of God. The demonstration, on the other hand, becomes possible only when that intellect is later granted such knowledge—that is, in the beatific vision. According to Ockham's own theory, such action on the part of God can happen only "supernaturally," that is, not in the normal course of nature. A clear case of this type of demonstration would be if I now doubt the existence of God. When I die, I may be lucky enough to enjoy the beatific vision. As blessed, I no longer doubt the proposition 'God exists'. I can then prove this proposition using my newly gained intuitive knowledge. This would satisfy the requirement that the proposition first be doubted and would, at the same time, satisfy the requirements of "evidence" that are required for demonstration.

Leaving aside the intricacies of Ockham's theory of demonstration,[25] it is clear that Ockham and Aquinas have radically different notions of just what falls within the scope of a science of theology. Indeed, Ockham leaves us with only one kind of syllogism—those in which a connotative or negative concept that is common to God and creatures is predicated of the divine essence—that, properly speaking, is a scientific demonstration. What is also clear, however, is that such propositions are not, strictly speaking, theological, but fall, rather, within the scope of metaphysics. Ockham himself never asks the question that occupied Aquinas's interest: Whether, outside the philosophical disciplines, some other doctrine is required. For Ockham, it is clear that when one talks of salvation, certainly such doctrine is required. However, that doctrine, according to the criteria just offered, is most definitely not a science.

Furthermore, the propositions that are demonstrable offer no specific knowledge of God. For the very example Ockham uses would apply to any being: 'All beings are good, Brownie the Donkey is a being, therefore Brownie is good'. Ockham leaves no room, either in theory or in practice, for a science of theology that offers any insight into the nature of God as God. The result of Ockham's position concerning theology is twofold. On the one hand, he removes metaphysical causality as the foundation of science and replaces it with epistemic causality in the notion of evidence. On the other hand, and perhaps as a consequence, he removes God from any metaphysical relation to things that could form the basis of science. As a result, science is no longer able to reach behind the existing singulars to uncover some sort of rational ground. Knowledge of the things is unable to penetrate the order of genesis. Science must now confront the sheer givenness of things and secure its epistemology in the face of that givenness. Ockham's task is to ground *scientia* without having recourse to the metaphysics of causality—that is, without recourse to the rational ground of existing things.

Singulars, Intuitive Knowledge, and Evident Knowledge

There is, perhaps, no issue more hotly contested in the literature on Ockham than his theory of intuitive knowledge.[26] The main contributions in the debates of modern interpreters have focused on the question of skepticism as a consequence of Ockham's theory.[27] This question will not be the main focus of our treatment here, though we shall treat it in due course. Rather, the focus here will be on the role that intuitive cognition plays in granting some knowledge the claim to be evident.

As we have seen in Ockham's definition of a proposition that is knowable by science, one of the main characteristics of such a proposition is that it is able to be made evident by propositions that are necessary and evident.[28] Working from this definition backward, we can ask what sorts of propositions are necessary and evident. Here Ockham diverges from the traditional account of such propositions. As we saw in Aquinas, and as holds for almost every other commentator on *Posterior Analytics,* such propositions must be *per se* according to the various modes in which a proposition can be said to be *per se.*[29]

Ockham, however, expands his concept of science to include premises that are not self-evident. The question, then, is how to ensure the evidence of these propositions, because, according to his definition of science, the conclusions of a scientific syllogism must be able to be made evident by propositions that are themselves evident. According to Ockham, the premises can be propositions that can be evidently known through experience.[30] As we have seen, Ockham goes so far as to claim that some sciences have principles that can only be grasped by experience and are in no way self-evident:

> There are many sciences which are not able to be acquired without experience, although opinion and credulity concerning these propositions which are knowable is able to be acquired through precise teaching. Therefore,

those sciences do not have self-evident principles, others, however, have self evident principles.

[. . . multae sunt quae non possunt adquiri sine experientia, quamvis opinio et credulitas de illis scibilibus possit adquiri per doctrinam praecise. Igitur illae scientiae non habent principia per se nota, aliae autem habent.][31]

From this, two problems immediately follow: (1) if experience can lead to principles that are evident, though not self-evident, what kind of experience must be had in order to ensure that evidence; and (2) if our experience is only of contingent singulars, how can we derive propositions that are universal and necessary.

Apprehension and Judgment: The Context of Intuitive Knowledge

Ockham begins his treatment of intuitive knowledge from a division of acts of the intellect. He states that among all the acts of the intellect, we can distinguish two that, while remaining distinct, have a special relationship. On the one hand, there is the apprehensive act of the intellect. This act, according to Ockham, concerns whatever is able to complete the act of the intellective power. This refers both to terms and what they signify as well as propositions and what they signify.[32] On the other hand, there is the judicative act. Through this act the intellect not only apprehends its object, but in addition assents to it or dissents from it. This act, unlike the apprehensive act, is only with respect to propositions. Assent and dissent are related to truth. When we estimate that a proposition is true, we assent to it; when we estimate that it is false, we dissent from it. Truth, in turn, is a property that belongs only to propositions.[33]

The relation between these two acts is that the judicative act of some proposition (or what is signified by that proposition) presupposes an apprehensive act of the same proposition (or what is signified by that proposition).[34] Furthermore, there is a relation between the judicative act with regard to a proposition and an apprehensive act with regard to the terms that make up that proposition: " . . . every judicative act presupposes in the same power [i.e., in the intellective power] simple knowledge of the terms, because it presupposes an apprehensive act. And an apprehensive act with respect to some proposition presupposes simple knowledge of the terms" [. . . actus iudicativus praesupponit in eadem potentia notitia incomplexam terminorum, quia praesupponit actum apprehensivum. Et actus apprehensivus respectu alicuius complexi praesupponit notitiam incomplexam terminorum].[35] In other words, the forming of a proposition, which is required for the judicative act, requires the grasping of the simple terms involved in the proposition. These simple terms have singulars as their significata.[36] Thus, the act of assenting to or dissenting from a proposition requires the grasping of the singular terms from which that proposition is composed.

Here, however, a caution must be raised. The term "apprehension" seems to carry with it a connotation that this grasping is sensory. Ockham distinguishes between sensory apprehension and intellectual apprehension. However, sensory apprehension is not the immediate cause of the judicative act of the intellect. For this, there is required an intellectual apprehensive act. This act, then, is the immediate proximate cause of the judicative act of the intellect.[37] This is not to say, however, that a sensitive, apprehensive act has no role to play at all. Quite the contrary, where the object in question is sensible, then sensitive apprehension is required, though it is not enough. What is required, in addition to this apprehension, is an intellectual apprehension of the same object. It is this latter apprehension that would be the *proximate,* immediate cause of the judicative act. When the truth in question is a contingent truth, then a sensitive, apprehensive act is required—in this life—along with and in addition to the intellective act.[38]

Ockham and Aquinas on Apprehension and Judgment

The context of this distinction between apprehensive and judicative acts of the intellect is not difficult to see. The problem that Ockham faces here is precisely that which Aquinas faced in his *prooemium* to *Posterior Analytics.* The problem, as Aquinas had phrased it, was that science is ratiocinative, that is, it moves from the knowledge of one thing to the knowledge of another thing. Therefore, the knowledge of the first thing must be secured already before I can begin a demonstration and move to the knowledge of the second thing.

To solve this problem, Aquinas developed a threefold distinction among acts of the intellect: grasping simples (terms and their significata); grasping complexes (propositions and the states of affairs that they assert); and judgment or moving from the knowledge of one thing to the knowledge of another. What Aquinas there argued was that the first two acts of the intellect apprehend either simple things or complexes in such a way that once I grasp them in this way I can immediately know any proposition or state of affairs that is based on them. My ability to grasp and then to know with certainty a proposition is required if Aristotle's ideal of science is to be met.[39] Ockham's treatment here falls completely within that context. He mirrors Aquinas's language to such an extent that it is difficult to see that he did not have in mind here a sort of *prooemium* to *Posterior Analytics.*[40] Moreover, not only does he mirror that language, he does not stray all that far from the position Aquinas achieved in his own commentary.

We can now complete at least one part of the story that began in Aquinas: the *Prinzipienproblem*. Aquinas, as we saw, realized immediately that the main source of difficulty surrounding a would-be science of theology is that there is no *intellectus* of its principles because that is precisely what is to be expected in the beatific vision.[41] Rather than make an argument about how we can have *intellectus* of God, Aquinas chose to make theology a science that is subalternated to those sciences that precisely do have such *intellectus*. Aquinas appealed to the *intellectus* that comes in the beatific vision. Therefore, when Ockham raises the issue of *notitia intuitiva,* he appeals to a context as well as a concept that is part of the discussion of *intellectus* in commentaries on Aristotle's *Posterior Analytics*. This point has not been stressed in the literature.[42] This will be important when we turn to the discussion of skepticism below. It is, however, within this context that we have to assess Ockham's position.

The main difference between Ockham and Aquinas is the role of judgment. While Aquinas also posited apprehension and judgment, for Aquinas judgment arises only in that part of logic that deals with ratiocination, that is, with science.[43] For Aquinas, the apprehensive act applies both to simples as well as to complexes, as it does for Ockham. However, the assent to the truth of a proposition is not called *judgment*. Aquinas reserves the term "judgment" for ratiocination, that is, for the "certitude of science." This means that the grasping of a proposition and the assent to it comes without the act of judging. Assent must follow directly from the type of grasping with which I apprehend the proposition. Apprehension and assent must be so tightly connected that it does not make sense to separate them.

For Ockham, judgment occurs already before the act of ratiocination. Ockham notices that I can apprehend a proposition in the ways spelled out by Aquinas (i.e., that I know the "*quid est res*") and yet the intellect can withhold assent.[44] If one act can be without the other, they must be separate acts. This is the source of Ockham's problems. For if the intellect can withhold assent on this level, the level of grasping propositions, then the intellect must have some grounds for assenting to a proposition or dissenting from it. This leaves room for error at a fundamental level of the theory of science, the level of *intellectus*. Aquinas did not hold out the possibility of judgment on this level, and, therefore, the level at which error could occur was postponed to the level of ratiocination, that is, of moving from the knowledge of one thing to the knowledge of another. For Aquinas, error in science is merely a logical problem, for Ockham it is epistemological.[45]

This means that for Aquinas the whole question of certitude in judgment rests on two factors: (1) the original apprehension of simples and the complexes (propositions/states of affairs) made from those simples; (2) the rules of syllogistic logic that make sure that in going from the knowledge of one thing to the knowledge of another thing, there is no "defect" in the truth. Thus, science, if it is to have certainty, must "resolve" judgment to its material principles, that is, the propositions from which it is built. Science must always be sure that it begins with the kind of apprehension that would know the "*quid est res.*" Once the resolution has been performed, what remains is to follow the rules of syllogistic logic to ensure that no error has crept in.

The problem of error is more fundamental in Ockham because it can take place on the level where *nous* (*intellectus*) is to take place. This is not to say that *notitia intuitiva* is *nous* for Ockham. Rather, just as for Aquinas, *nous* requires first an apprehension of simples, second an apprehension of propositions composed of those simples, and finally the formation of propositions that are universal and necessary, yet which are still grounded on the original apprehension of simples.[46] While for Aquinas, judgment, and, therefore, the possibility of error arises only after this process has taken place, for Ockham judgment takes place before I have achieved *nous*.

This difference in the site and role of judgment points to a more serious and penetrating difference in the relation between rational ground (reasons) and existing singular. For Aquinas, the "*quid est res,*" the quiddity, is something that is given to the intellect in an apprehensive act. This means that the existing singular is already apprehended in terms of its essence—it is always a particular of some universal. Therefore, as we have seen, the apprehension of the thing is already essential, that is, what is grasped is already the universal nature of the thing. It is difficult to see how Aquinas can have a meaningful notion of existence at all.[47] The grasping of the existing singular is such that it always comes *through* the universal nature. It is this universal nature that provides the ground for the existence of the singular. The singular exists always *as* something more universal.

For Ockham, conversely, judgment happens at the point at which the intellect moves from the existing singular to making any claims about it whatsoever—including positing a "universal nature." The act of assenting or dissenting—which is the definition of judgment in Ockham—arises when one forms a proposition on the basis of an apprehensive act. The apprehensive act grasps, for instance, Socrates, whiteness, the proposition 'Socrates is white', and even an entire demonstration.[48] The judicative act, on the other hand, does not pertain to particulars, but to "complexes" (i.e., facts and propositions asserting these facts). The judicative act assents to or dissents from a complex. If Socrates is not white, a judicative act can assent to the complex 'Socrates is white' or dissent from it, or be withheld. Thus, the judicative act can be right or wrong. I could grasp Socrates and yet dissent from the complex 'Socrates is human'. This must mean that the rational activity of grasping universal natures always comes after the apprehension of the thing and this activity could go awry. The universal nature, therefore, is not the ground for Socrates's existence, nor for my apprehension of Socrates. Rather, the assertion that Socrates is human remains an activity of the soul that is grounded *in* the soul's knowledge of the existing singular—it is not the ground *of* the existing singular. This is the result of Ockham's move from metaphysical to epistemological causality.

Since the level at which error can enter is pushed back to a more fundamental one in Ockham, a way to avoid that error is similarly pushed back to that level. The distinction between intuitive and abstractive cognition, therefore, is to provide a means for correcting the error at this fundamental level. While for Aquinas the guide for correcting error in judgment was the rules of logic, for Ockham the guide is whether my original apprehension of the simples is intuitive or abstractive. If my original apprehension is of the intuitive sort, then the *intellectus* that results from that apprehension will be certain and the scientific syllogism I construct from those propositions grasped by *intellectus* will be certain as well. The issue of skepticism arises at precisely this point: How can I ever be sure that my original apprehension was intuitive? We will return to this point below when we take up the question of skepticism directly.

Two Kinds of Apprehension: Intuitive and Abstractive

The two forms of apprehension we have been discussing, that is, sensitive and intellective, can be further divided into two forms: intuitive cognition and abstractive cognition. Ockham uses the distinction between intuitive and abstractive cognition to prove that there is a distinction between the apprehensive act and the judicative act. Ockham begins by distinguishing two sorts of simple knowledge of terms:

> Every simple knowledge of some terms which is able to be the cause of evident knowledge with respect to some proposition composed of those terms is distinguished by species from simple knowledge of those terms which, however much it is intended, is not able to be the cause of evident knowledge with respect to the same proposition.

> [Omnis notitia incomplexa aliquorum terminorum quae potest esse causa notitiae evidentis respectu propositionis compositae ex illis terminis distinquitur secundum speciem a notitia incomplexa illorum, quae quantumcumque intendatur non potest esse causa notitiae evidentis respectu propositionis eiusdem.][49]

Here, the distinction is between an apprehension that causes evident knowledge of some proposition and an apprehension that is not able to cause evident knowledge of the same proposition. Ockham concludes the argument by stating that if I can apprehend some terms, and yet not know evidently whether a proposition composed of those terms is true or false, this would prove that the apprehensive and judicative acts are indeed distinct.[50] Thus, Ockham ties the judicative act of the intellect to evident knowledge: Only when I have an apprehensive act that causes evident knowledge of some proposition can I then judge that proposition, that is, assent to it or dissent from it. When Ockham turns to his discussion of abstractive and intuitive knowledge, the same distinguishing feature is used to separate them:

> Abstractive knowledge, however, is that by virtue of which it is not able to be evidently known, concerning contingent things, whether they are or are not. And in the same way, abstractive knowledge abstracts from existence and non existence, because through abstractive knowledge it is not able to be known evidently that an existing thing exists, nor that a nonexisting thing does not exist, in contradistinction to intuitive knowledge.

> [Notitia autem abstractiva est illa virtute cuius de re contingente non potest sciri evidenter utrum sit vel non sit. Et per istum modum notitia abstractiva abstrahit ab existentia et non existentia, quia nec per ipsum potest evidenter sciri de re existente quod existit, nec de non existente quod non existit, per oppositum ad notitiam intuitivam.][51]

The distinguishing characteristic that separates intuitive from abstractive knowledge is that one is able to be the cause of a judgment and the other is not. This is the meaning of "evident knowledge."[52] If some knowledge is able to ground a judicative act, that knowledge is called evident. What should be noticed, however, is that these judgments include, but are by no means limited to, existential judgments.

What kind of knowledge, then, can lead to a judgment? Ockham calls this kind of knowledge "intuitive." It is, indeed, *intellective* intuitive knowledge that is able to be the cause of a judgment: " . . . it is seen that only intellective knowledge is sufficient for a judgment as a proximate cause, and that the intellect has cognition of time and other such things just as the senses. But this is not able to be without intuitive knowledge . . ." [Ex istis auctoritatibus videtur quod sola notitia intellectiva sufficit ad iudicum tamquam causa proxima, et quod intellectus ita habet cognoscere tempus et huiusmodi sicut sensus. Sed hoc non potest esse sine notitia intuitiva . . .].[53] These two forms of intuitive knowledge, that is, sensitive and intellective, are distinct, according to Ockham. Yet, such contingent truths are not able to be known concerning these sensible things unless when they fall under the senses, because intellective, intuitive knowledge of these sensible things is not able to be had, in this life, without sensitive, intuitive knowledge of them.

[tales veritates contingentes non possunt sciri de istis sensibilibus nisi quando sunt sub sensu, quia notitia intuitiva intellectiva istorum sensibilium pro statu isto non potest haberi sine notitia intuitiva sensitiva eorum.][54]

It is clear that evident knowledge can pertain to contingent truths—whether these truths belong to the intellect only, or, in addition, originally fell under the senses. For us, in this life, when the truth under consideration is one that comes from the sensory world, then sensitive, intuitive knowledge is required for intellective, intuitive knowledge, which, in turn, is able to ground a judgment.

This intuitive knowledge is that knowledge, "by virtue of which it is able to be known whether the thing is or is not."[55] Therefore, if we have intuitive knowledge of a thing, and the thing is, the intellect immediately judges that the thing is and evidently knows that the thing is, "unless it be impeded by some imperfection of that knowledge."[56] Furthermore, and more importantly, intuitive knowledge grasps whether something inheres in something else, whether it is in a spatial relation to something else, and, in general, whether it is related in some way to something else. From this intuitive knowledge it is known immediately that such is the case.[57] The consequence is that such intuitive knowledge would lead immediately to a judicative act, that is, to assenting to a proposition such as "Socrates is white," if Socrates is, in fact, white. It "leads to" a judgment because it is the cause of evident knowledge of that proposition, but it is not the cause of the judgment. The will alone is the cause of assent or dissent.

Intuitive Knowledge and the Issue of Skepticism

The issue here is whether one has sufficient knowledge of a thing or a state of affairs so that one can form a judgment about that thing or state of affairs and not go wrong.[58] It is in this sense that the charge of skepticism enters the discussion of Ockham's theory. In its mildest form, the charge is one of circularity:

> If Ockham assumed that, because false judgments ensue, the initial non-veridical apprehensions were not intuitive, a range of problems remains. If he has not thereby contradicted the priority of intuitive cognition, he has shifted the difficulty from judgment to that of providing some criterion for introspective discrimination of intuitive and abstractive cognitions. Such an

assumption is further objectionable insofar as it makes the nature of what is causally prior (apprehension) dependent upon its effect (judgment).[59]

The problem is that Ockham asserts that the judicative act presupposes the apprehensive act and that the apprehensive act, when it is intuitive cognition, is the cause of the judgment. However, when a judgment is wrong, the only answer Ockham can give is that the apprehensive act was not intuitive (or that it was impeded in some way). Therefore, in order to determine whether a given apprehensive act is intuitive or not, one seems to have to wait until a judgment is made and then determine whether that judgment is true or false, that is, whether the state of affairs is as the proposition says it is. Consequently, the nature of intuitive knowledge seems to depend on what is posterior to it, that is, the judgment that is dependent upon that apprehensive act.

This same problem is highlighted by T. K. Scott:

> [Ockham] seems convinced that there must be instances of intuitive cognition, evident judgment and necessary judgment, if a science is to be possible. But at the same time his discussion of the interrelations among them depicts them as internally related in such way that any attempt to determine what is and what is not an instance of any one of them involves a circularity that makes the determination impossible.[60]

The circle begins immediately with the definition of intuitive knowledge given above, namely, that knowledge by which one is able to judge that a state of affairs is in reality as it is in some proposition and by which one is able to know that proposition evidently.[61] The very definition of intuitive knowledge is that it causes an evident judgment. When one then turns to the definition of an evident judgment, one finds it defined as:

> Cognition of some true complex [i.e., proposition], which is of such a nature that it is caused sufficiently, either immediately or through some mediation, by simple knowledge of the terms. Thus it is that when simple knowledge of some terms, whether they are terms of that proposition or another proposition or diverse propositions, sufficiently causes in any intellect having such knowledge, either immediately or through some mediation, knowledge of a complex, then that complex is known evidently.

> [notitia evidens est cognitio alicuius veri complexi, ex notitia terminorum incomplexa immediate vel mediate nata sufficienter causari. Ita scilicet quod quando notitia incomplexa aliquorum terminorum sive sint termini illius propositionis sive alterius sive diversarum propositionum in quocumque intellectu habente talem notitiam sufficienter causat vel est nata

causare mediate vel immediate notitiam complexi, tunc illud complexum
evidenter cognoscitur.][62]

This simple knowledge of terms can be nothing other than intuitive cognition. Therefore, the definition of evident knowledge is one that is of a
nature to be caused by intuitive cognition, while intuitive cognition is
defined as that which is able to cause evident cognition. The circle is
complete!

That this leads to skepticism is apparently a consequence Ockham
never saw, and, indeed, was unwilling to countenance. Ockham repeatedly
refuses validity to any skeptical arguments—that is, to arguments that hold
that we are unable to know anything with certainty. How do skeptical
consequences follow? If certitude is based on intuitive cognition in order
to be certain that I know anything concerning the external world, I must
be able to pick out which of my cognitions are intuitive and which are
not. However, the circle that Ockham gives relating evident knowledge to
intuitive knowledge and intuitive knowledge to evident knowledge pre-
vents my picking out which of my cognitions are intuitive. Therefore, I am
never able to know which of the propositions I know at any given time
are certain and which are not. Therefore, I am never able to know with
certitude that I know anything about the external world.[63]

The skeptical argument depends upon two theses to which Ockham
would never agree. First, it depends upon the positing of a dichotomy between subject and object that is foreign to medieval philosophy.[64] Second,
it depends upon the thesis that certain knowledge is, in fact, possible. When
one attacks medieval thinkers with the charge of skepticism, neither of
these theses are demonstrated. Yet it is the validity of these positions that
Ockham's epistemology itself calls into question. It is at this point where
we see the real impact and power of Ockham's epistemology.

Ockham stands on the threshold of subjectivity. The theory of knowledge that provides the rational ground for the existing singular was not
threatened by skepticism because the order of coming to know retraces the
order of coming to be. Once, however, the order of coming to be is seen
always to exceed and therefore, to some extent, always to stand outside reason, knowledge becomes threatened. But the knowledge that is threatened
is only that knowledge whose goal is the rational ground of existing singulars. This means that the existing singular must present itself—must an-
nounce itself—as always standing outside the knowing soul.[65] This
characteristic of standing outside and announcing itself does not yet mean
that the object stands over-against a subject whose existence is already secure. Rather, it means that the object announcing itself and the soul to
which it announces are by their very natures connected.

Ockham's position on intuitive knowledge leaves him in a somewhat precarious position. For on the one hand, he loosens the tight connection that medieval philosophy held between the soul and the world. Ockham's insistence on the existence solely of individuals means that the link between the conceptual realm and the realm of existing things is a tenuous one—held together only through the notion of natural signification. On the other hand he is not yet willing to argue that there is no natural connection. He stands, then, as it were on the threshold of the subject, unwilling to break completely the ties that the soul has with the world. On the other hand, his insistence on the absolute power of God further loosens the links between soul and world by denying a natural theology. He comes the closest of any medieval philosopher to positing the subject/object split. Perhaps this is why he is so often labeled a skeptic.

In his distinction between intuitive and abstractive knowledge, then, Ockham differs from Aquinas in only two ways: (1) he has a more sophisticated understanding of the ways in which we can grasp terms and what they signify; (2) he moves the possibility of error to an earlier stage because of his understanding of the grasping of terms and the formation of propositions.[66] Furthermore, Ockham follows Aquinas in arguing that *intellectus* begins with singular, contingent propositions. As has been pointed out by Stephen Dumont, the innovation in Scotus's distinction between intuitive and abstractive knowledge was not the intuitive side of the distinction, but rather the abstractive side.[67] Abstractive cognition in Ockham comes to be a grasp of simple things and terms such that I can only grasp those (conceptual) features that belong to the thing independently of its existence in its contingent relations in the world.

If Ockham's procedure is open to error and therefore open to skepticism, it is no less so than Aquinas's. For Aquinas could only prevent skepticism by appeal to the order of the cosmos—that is, to natural theology—which reason can follow so as to avoid error. The order of the cosmos itself, however, is never grasped with any certainty, but only in revelation. What must be understood is that for all medieval thinkers who paid attention to *Posterior Analytics,* the entire logical system of Aristotle was always understood as predicated on the uniquely human ability to err. That any system of logic and science could get rid of error completely would be miraculous. It seems clear that the problem of cognitive certainty in terms of complete removal from error was not an issue for medieval thinkers. To get rid of error would, in a sense, repeat our original sin: We would have the knowledge that God has.

CHAPTER 6

THEOLOGY BEYOND SCIENCE

The Separation of Theology and Science

We have already seen how Ockham limits those theological propositions that are knowable by science. It is not yet clear, however, that this is the result of a complete break between theology and science. Science, for Ockham, is no longer theocentric. But does that fact mean that theology is not a science?

Ockham's rejection of four of the six classes of theological propositions as scientific hinges, as we saw above, on the question of demonstrability. His overriding concern is to show that most theological propositions do not fit the formal, logical criteria for demonstration. Indeed most of the rejected classes are rejected because of the impossibility of finding a middle term that would be prior to both God and the property being demonstrated about God. The two accepted classes of theological proposition are accepted precisely on the grounds that a middle term is available to us through which some concept can be demonstrated of God.

The inability for us to achieve a proper middle is one of the central problems in assigning theology, *tout court,* the name *scientia.* Whatever else it may be, and whatever else its merits, theology, in large part, consists of propositions that are simply indemonstrable. The issue in our inability to find a proper middle is the divine simplicity. There is no property of God that is not itself essentially God. We cannot artificially separate parts of the divine essence and use them to demonstrate other properties. Whatever God is, God is immediately and essentially. Where can we find the middle term that would come between one of God's properties and the divine essence?

In those cases in which we can find a middle term, we are left with propositions that are almost entirely vacuous. We are left with propositions in which some negative or connotative concept that is common to both

God and creatures is predicated of the divine essence. As we saw, the example Ockham gives is, "'Every being is good; God is a being; therefore God is good.'"[1] The problem here is that the demonstration works just as well when some other being is substituted for God: 'Every being is good; Brownie the donkey is a being; therefore Brownie the donkey is good'. This latter demonstration is just as vacuous as the former.

The other central problem raised by theological propositions is their dubitability. We have seen that this is one of the characteristics of a proposition that is knowable by science. Yet, several classes of theological propositions will not be knowable by science because they cannot be doubted. When we look for concepts that are common to God and to creatures, we run into precisely this problem. We have just seen how some of these concepts can be demonstrated, though vacuously. The determining factor was that they were connotative, and therefore the subject and predicate terms stood in for (and indeed signified) the same thing: Being signifies some thing and what is good is a being that is desirable—both terms signify the same thing, though good consignifies something in addition to that. What makes these propositions demonstrable is that the common concept is not predicated essentially of God. God is good because God is a being and not because God is God. We simply cannot demonstrate that God as God is good—although God as God is good. When we attempt to predicate common concepts essentially of God and creatures, the proposition turns out to be immediate, that is, *per se nota,* though not for us in this life.

Ockham has one other class of theological propositions whose propositions are, in principle, demonstrable. This class includes those propositions that are dubitable for us and that predicate a common composed concept that is proper to God. These propositions are demonstrable using the divine essence, some distinct cognition of deity, or something common as a medium. However, these propositions are demonstrable only in principle. On the one hand, they satisfy the criterion of dubitability, yet, on the other, they cannot be demonstrated by those who doubt it. A proposition like 'God exists' is such a proposition. It is possible for us, in this life, to doubt this proposition. Yet we ourselves cannot demonstrate it. It only becomes demonstrable when "God causes intuitive or abstractive knowledge" in the intellect of the one who first doubted the proposition. Now, according to Ockham's own position, such a person, in whom God caused intuitive knowledge of the deity, is not a *viator*—that is, it is not possible in this life. For, on the one hand, if God causes intuitive knowledge of the deity through God's absolute power, then that person is either blessed (i.e., enjoying the beatific vision) or damned.[2]

Here Ockham differs from Scotus. While Scotus thought that abstractive knowledge of deity was enough to ensure that theology is science, for

Ockham it is not enough. They agree on the fact that abstractive knowledge of God is possible in this life without violating the *viator* state—though whether this is possible only *de potentia absoluta* or not is a question. For Ockham, abstractive knowledge of deity is possible *de potentia ordinata,* yet it is never enough to cause evident knowledge of some conclusion—the very requirement for science.

As a consequence, abstractive knowledge is not of such a nature to produce the evidence required for this type of syllogism in question. For science requires that the conclusion be made known through a syllogism whose premises are themselves evident. Ockham is clear that abstractive knowledge cannot cause evident knowledge—this is the very criterion by which it is distinguished from intuitive knowledge. In short, Ockham's theory of science is a theory in which evidence is passed along from the premises to the conclusion. Ockham states that one who doubts "God exists," can later be granted intuitive or abstractive knowledge through which one may prove that conclusion: The divine essence is; God is the divine essence; therefore, God is.[3] The issue is what kind of knowledge would provide evidence of the proposition "the divine essence is." It must be intuitive knowledge, and this for two reasons. First, and foremost, I know of no place where Ockham allows abstractive knowledge as knowledge that causes evident knowledge. That role is restricted to intuitive knowledge. Even if abstractive knowledge is somehow involved, there must be at the ground some intuitive knowledge. Ockham argues that abstractive knowledge itself is caused by intuitive knowledge, if only partially. Secondly, the proposition 'The divine essence is' is obviously an existential proposition. It is clear, however, that only intuitive knowledge can cause evident knowledge of existence. Therefore, the syllogism that Ockham proposes is one that is scientific *only for the blessed.*

Here again, however, we have a vacuous sort of scientific proposition. For the blessed are distinguished from the *viatores* precisely because they have intuitive knowledge of deity, and that must include, according to the definition of intuitive knowledge, knowledge of existence.[4] Therefore, the proof of the existence of God is entirely superfluous for those who already possess evident knowledge of God's existence. It would be, for Ockham, like trying to prove the existence of Socrates while Socrates is standing in front of you.

In short, this class of theological propositions that are demonstrable are never demonstrable for us, but only for the blessed, that is, those already enjoying the beatific vision and for whom such demonstrations are superfluous. Furthermore, Ockham here specifically cuts off any relation between the blessed and the *viatores,* thereby leaving us without a science of theology. We see here that Ockham is operating in exactly the same milieu as was Aquinas—the evident knowledge that is required in order for theology to be a science is lacking for us, but is possessed by the blessed. Ockham, in turn, rejects the option that our science of theology is subalternated to the science of theology that the blessed and God possess.

Ockham could see the benefits of the Aquinian solution—that it would provide premises that are evidently known so that one could formulate a theological syllogism that would be scientific—and yet rejects it. His rejection follows from his close reading of the argument. He does not accept anything like Aquinas's analogy and sees that the principles that would come from the science of God and the blessed would be merely believed. A syllogism that has premises that are believed will produce a conclusion that is also merely believed.[5]

For Ockham, theology is, for the most part, an "inevident veridical habit." The rest, however, is not a veridical habit because it is not judicative but apprehensive.[6] The fact that it is either inevident or apprehensive means that it is not scientific—for science is an evident, veridical habit. What are the consequences, then, of this rejection of the Aquinian solution?

The overriding consequence is that science, in general, can no longer be centered around God's science. There results an epistemological separation between the creator and creature. Our science cannot look to the divine knowledge of creatures as a measure of our own knowledge. Our knowledge and God's knowledge do not terminate at the things—or, if they do, that cannot be the ground of our epistemology. If there is to be a science at all, then that science must account for itself, in the world, and not appeal to God as its guarantor. Finally, because we have no access to the divine science, any appeal to the order of the universe with its necessary laws to ground the necessity of our science is rejected. With this separation, science can no longer function as the rational ground of existing singulars.

This separation leads in two directions. First, it leads in the theological direction. One finds no natural theology in Ockham. Rather, theology is a habit of faith and, as such, must remain with the God of faith. The God of faith, then, is the God who is omnipotent, omniscient, savior, simple, and unlimited. God's activity is no longer placed in the constraints of natural law—which inevitably operated as a law instituted by God but also, and more seriously for later thinkers, as a law making God's actions intelligible and, therefore, restraining God. This is not to say, however, that this concept of God separated from science has no impact on science. Quite the contrary, the newly freed characteristics of this God—especially omnipotence—have far reaching effects on science, especially the science of nature.

God's Omnipotence and Science

According to Ockham (who in some ways follows tradition and in other ways departs from it) God's power is one and simple.[7] However, we can distinguish our experience of that power in two ways. These two ways of viewing the power of God are by no means contrary. Ockham defines them as, on the one hand, God's power to do something according to the laws that God has already instituted. God's absolute power, on the other hand, is the power to do *anything,* as long is it does not include (or lead to) a contradiction.

An enormous interpretative tradition was built up around this distinction. In the secondary literature roughly before 1930, it was seen that theologians of the fourteenth century put this distinction to use for the destruction of the synthesis that had been achieved, most notably by Aquinas. On this reading, attention to God's absolute power led directly to idle speculations and it destroyed the relationship between God and the world. Most seriously, it led to skepticism. What is more, not only did it lead to skepticism, but those very skeptics flaunted and reveled in their skepticism.[8]

It now seems to be the consensus that Courtenay is correct in saying that the distinction is one that attempts to uncover the difference between the initial possibilities open to God versus those things that, out of all those possibilities, God has, in fact, chosen to create.[9] Ockham is clear that, since these are not two powers, God cannot act inordinately.[10] Ockham understands this to mean that of all the possibilities initially open to God, God chose this order through God's will. This order has no necessity, since God could have chosen an equally good order, yet one that is completely different. What it does not mean, however, is that God can violate that order at any moment. Nor does it mean that God could replace the order that has been instituted by a new and different order. I know of no passage in Ockham's works where he entertains such a possibility. Rather, since God has chosen to create this order, then things (both in terms of nature and in terms of grace) will function according to this order.

Therefore, the order of the world is contingent, not in the sense that it could be different, but in the sense that it *could have been* different. This makes all the difference in the world! The contingency of our world is a result of the fact that God had more possibilities open to the divine will than just this present order. The fact that *this* order was created is contingent because it derives only from the fact that God has chosen this order. God could still be God and yet the order of the universe (both in terms of nature and in terms of grace) could have been different. The contingency of the world, then, results from its relation to the divine will that is not bound by anything outside of the principle of contradiction.

God's Absolute Power and Natural Theology

The result of this is that one cannot use the present order to infer anything about God—except, perhaps, that God is creator, and, as we shall see, conservator. Courtenay has argued that one can "no longer accept, however, the thesis that Ockham . . . disallowed natural theology in principle."[11] Whether or not Ockham ever explicitly denied natural theology is, of course, not the issue here. The issue, rather, is whether one can accept this view of God's power and still, at the same time, admit a natural theology. On that score, however, certainly Aquinas, who assumes the same general principles, ought not to have had a natural theology either. And yet he did. Ockham, who does not explicitly deny a natural theology, does not really produce one either.

However, as we have seen, Aquinas does not always take his own principles very seriously.[12] The systematic instability created by God's absolute power, which Aquinas accepts in almost the same way as Ockham, was never brought to its completion in Aquinas. Rather, by prioritizing the divine intellect over the divine will, and furthermore, by making rationality fundamentally practical, he was able to overcome these instabilities and link, in a necessary way, God to the order God has created. This is precisely the move that Ockham wants to avoid. God is not part of the universe.[13]

Ockham makes only two gestures toward a natural theology. The first is that he allows that concepts that are connotative or negative and are common to God and creatures are able to be formed and demonstrated of the divine essence.[14] The second is that he allows a proof the for the existence of God based on the conservation of things in the universe.[15] However, this proof is, by Ockham's own admission, probable and is not a demonstration and therefore ought not to be called a natural theology. What is interesting about the argument that Ockham gives is that it proceeds not from production, but conservation. Ockham argues that this proof is more "evident" than a proof from production, such as the prime-mover argument.[16] The main reason Ockham chooses this argument over others is because while an efficient cause can cease to exist while its effect remains, a conservator must always exist while that which it is conserving exists.

However, by not extending a proof for the existence of God back into the production of the universe, that is, to creation, Ockham has also circumvented the associated problem of linking the divine intellect to the universe in such a way that God could not have created the world otherwise, or, for that matter, many worlds. In short, this argument provides no insight into the divine intellect at the time of creation. Furthermore, Ockham refuses to accept those arguments that are related to causality. Without a demonstration of or through divine causality, the link between knowledge of a thing and its causes is broken. It is no accident that knowledge of causes, of rational ground, is not posited in his definition of science. Two things must be noticed here. First, Ockham does not call this a *demonstratio* but an *argumentum*. The existence of God is not proven in this way, but it becomes "more evident" than in the proof that would attempt to prove a first efficient cause. Secondly, Ockham highlights the fact that such proofs perhaps ought not to be called "theological" because there is nothing specific to them that one without faith could not also have.[17] The only thing that separates theology from other habits is faith. Therefore, a proper concept of theology is one that begins from faith. In this way, then, God is indeed cut off from the present order in a way that only by faith can we appeal to God. The present order shows us only God's will to choose an order—it does not take us inside the divine mind.

God's Absolute Power and Our Knowledge of the World

The link between the first question of Ockham's prologue to his *Commentary* on the *Sentences* of Peter Lombard, "whether it is possible for one in this life to have evident knowledge of theological truths," and the second question, "whether evident knowledge of theological truth is science properly speaking," is seen in Ockham's definition of science:

> I say that a proposition which is knowable by science, properly speaking, is a proposition which is necessary, dubitable, and such that it can be made evident by necessary, evident propositions being applied to it in a discursive syllogism.

> [dico quod propositio scibilis scientia proprie dicta est propositio necessaria, dubitabilis, nata fieri evidens per propositiones necessarias evidentes, per discursum syllogisticum applicatas ad ipsam.][18]

Thus, the question of science is the question of making something evident. The fact that the conclusion of a scientific syllogism is evident arises from the fact that the propositions from which it is derived are themselves evident.

The question of evident knowledge arises directly from the nature of a demonstrative syllogism. As we saw in Aquinas's treatment of science, a demonstrative syllogism should begin with premises that are self-evident [*per se notae*]. This characteristic insures that the conclusion of a syllogism will be necessarily true. However, if one follows this requirement strictly, the set of conclusions that can be proven scientifically will be relatively small. More serious than that, however, is the question of how the necessity of such self-evident premises is to be grounded.[19]

The problem faced by Ockham's theory of science stems from his ontological commitment to singulars and some of their qualities and nothing else. "Everything there is, is in its existence and essence contingent, the whole world is a contingent, free posit of God's free creation" [Alles, was ist, ist in seinem Dasein und Wesen kontingent, das Ganze der Welt ist eine kontingente, freie Setzung, Gottes freie Schöpfung"].[20] The singularity of everything that exists implies, it seems, its contingency. Indeed, for Ockham, to exist means to have the capability of existing without any connections to any other actually existing thing. This is the most frequent way in which Ockham deploys the notion of the absolute power of God. That which is one thing must be such that God could maintain its existence while destroying all things to which it happens to be connected. In the face of such an ontology, and the contingency it seems to imply, Ockham had an enormous problem in his theory of science.

The problem, stated briefly, is this: How can one formulate propositions that are true, necessary, and universal when everything that exists is singular and contingent (except for God, who is singular, yet not contingent)? Propositions in such a case need some ground for their certainty—a ground that must come from a source other than the simple meaning of the terms of the proposition. In this way, a universal self-evident proposition, for Ockham, tells us nothing, really, about the world. 'All humans are animals' means, in the end, 'If a human exists, then it is an animal'.[21] Ockham's solution to this puzzle not only grounds necessary, universal propositions in propositions about contingent singulars, but, at the same time, broadens the scope that science may have—if not changes its nature outright.

God's Absolute Power and the Contingency of the World

It should be obvious by now that the fact that everything is contingent except God and certain propositions is derived not from the fact that the present order is open to radical intervention by God, but is derived from the fact that God had many initial possibilities open. The fact that the present world was the one that was created follows merely from the divine will and not from any sort of determination either in the divine mind or outside.

There are here two contingencies that must be kept separate. First, there is a kind of contingency that Ockham never seems to have seriously considered, that is, the contingency in which the order of the world could radically change at any moment. This kind of psychotic contingency would make any knowledge impossible. Indeed, it would scarcely permit any life. This contingency would be the result of God interfering in the order that

God has already established. Second, there is the contingency in which the very order of the world is seen to be consequent only upon the divine will and upon no other determination. It is this kind of contingency that Ockham accepts (as does Scotus and as should Aquinas). There is all the difference in the world between these two kinds of contingencies. On the one hand, the world becomes chaotic. On the other, the world is ordered, and yet that very order cannot be grounded. The difference is between a world that is, in principle, unknowable, and a world that is knowable but not grounded—it is knowable only in itself. What is missing is the rationality of creation itself.[22] The knowability of things must not derive from the rationality of the cause of the world because their existence does not derive from this source. To refer creation to the absolute power of God is to refer it solely to the divine will. As a result, created things exist without a rational ground, but only with the non-ground of divine power.

Science without Existence

One main issue that was never directly thematized by Ockham was the question of existence. For the contingency of the world that is a result of Ockham's taking seriously the absolute power of God as a sphere of possibility results in existence being a "given." "Existence," for Ockham, and "thing" "signify the same thing."[23] Whereas Aquinas remains entirely perplexed by existence and must construct an enormous theoretical apparatus in order to grasp it, for Ockham it follows precisely from the contingency of the world. At least one consequence of this, however, is that Ockham cannot glean any theoretical promise from the fact that *esse* can be predicated of both God and creatures. The univocity of the predication, contrary to Aquinas's "analogy of being," is based on the fundamental givenness of existence. As we saw with the one class of propositions he considers demonstrable, the fact that existence can be predicated of both God and creatures leads us practically nowhere. *Esse* for Ockham tells us nothing more about the thing of which it is predicated.

This means, however, that the focus of many interpretations of Ockham's theory of intuitive cognition has been misplaced. Indeed, Ockham himself says that this kind of knowledge should grant us knowledge of existence or nonexistence—as we have seen. However, this never could have been the most important facet of this kind of knowledge for Ockham. If truth is defined by Ockham as a characteristic of propositions in which the two terms (i.e., subject and predicate) stand for the same thing,[24] then what Ockham needs from intuitive knowledge is the very knowledge that the predicate stands for [*supponere pro*] the same thing as the subject. This is exactly what Ockham delivers in his definition of intuitive knowledge:

Intuitive knowledge is such that when some things are known of which one inheres in the other or one is far away from the other or by some other way is related to the other, immediately, by virtue of that simple knowledge of those things, it is known if the thing inheres or does not inhere, if it is far away or not far away, and similarly concerning other contingent truths.

[Similiter, notitia intuitiva est talis quod quando aliquae res cognoscuntur quarum una inhaeret alteri vel una distat loco ab altera vel alio modo se habet ad alteram, statim virtute illius notitiae incomplexae illarum rerum scitur si res inhaeret vel non inhaeret, si distat vel non distat, et sic de aliis veritatibus contingentibus. . . .]²⁵

How does this compare with his first definition of intuitive knowledge?

Intuitive knowledge of a thing is such knowledge by virtue of which it is able to be known whether the thing is or is not, that if the thing is, immediately the intellect judges the that it is and evidently knows that it is, unless by chance it is impeded on account of an imperfection of that knowledge.

[Notitia intuitiva rei est talis notitia virtute cuius potest sciri utrum res sit vel non, ita quod si res sit, statim intellectus iudicat eam esse et evidenter cognoscit eam esse, nisi forte impediatur propter imperfectionem illius notitiae.]²⁶

The question comes down to how one interprets the statement: *eam esse* [that it is]. The formulation is originally found in Aristotle: "A falsity is a statement of that which is that it is not, or of that which is not that it is; and a truth is a statement of that which is that it is, or of that which is not that it is not."²⁷ It is also found in Anselm's *De Veritate* as the definition of truth as rectitude.²⁸ Clearly for both, but more in the case of Aristotle than of Anselm, the statements that can be true or false are not limited simply to statements of existence or nonexistence.²⁹ Aristotle, in the *Metaphysics* goes on to say that when I deny the statement 'Socrates is white', I am denying nothing else than the fact that whiteness belongs to Socrates. I do not deny, however, the existence of Socrates nor the existence of whiteness. I deny the existence of whiteness *in* Socrates. This definition of truth is translated into Latin with the same grammatical structure (infinite with an accusative subject) we have seen above.³⁰

If truth is not to be restricted to statements of existence, then the very structure we have been investigating must be seen in a new light. When Aristotle says that truth is a statement that says of something that is, "that it is" [*eam esse*] or of something that is not, "that it is not" [*eam non esse*], he is understood by Ockham to be making the claim that truth is saying 'X is Y' if X is, in fact, Y. 'Socrates is white' is true if and only if Socrates is white.[31] In short, this strange expression that is found in Aristotle, Anselm, and many others in the medieval tradition must be understood as a shortened, formal expression if it is to have any sense at all. What Ockham appears to have drawn from this Aristotelian definition of truth is that an existential statement (e.g., 'Socrates is' or 'Socrates exists') is by no means the most important case of a proposition that can be true or false, nor is it the determinative example for other propositions. It seems quite the contrary. Statements of the form 'X is Y' are the classic examples of propositions that can be true or false. A statement of the form 'X is' is only a special class of statements.

How, then, are statements of the form 'X is' to be interpreted? One hint Ockham gives us is that they are to be interpreted as a statement that includes a place-holder: 'X is . . .' where the " . . ." is to be filled in by a predicate term, turning the proposition into the classic example of a proposition that can be true or false, that is, of the form 'X is Y'.[32] Aristotle himself offers that interpretation in the Metaphysics. The definition of truth given in Aristotle's Metaphysics, that is, saying of that which is that it is and of that which is not that it is not, comes within a discussion of what constitutes contraries. What Aristotle considers to be the main feature of contraries is that it is impossible for them to belong to a subject at the same time.[33] In that sense, then, Aristotle is clearly referring to things that can belong to something else. The definition must be understood as a shorthand expression that includes a place-holder for a predicate that belongs (or does not belong) to a subject. There is no immediate existential import to the notion that truth is saying of something that is that it is.[34]

This makes sense of Ockham's notion of intuitive knowledge as well. For when Ockham goes on to interpret or explicate what is meant by judging of something that is, that it is, he immediately refers to predicates: the inherence of a property, the relation of space, and other relations between two terms. Once one notices this fact, one sees aspects of Ockham's theory that have not surfaced before. When Ockham goes on to talk about intuitive knowledge of a nonexistent object, his language is very precise:

> Therefore I say that intuitive knowledge and abstractive differ in themselves and not on account of the object, nor on account of whatever is their cause, although naturally intuitive knowledge is able to be without the existence of the thing, which is truly the mediate or immediate efficient cause of intuitive knowledge.
>
> [Ideo dico quod notitia intuitiva et abstractiva se ipsis differunt et non penes obiecta nec penes causas suas quascumque, quamvis naturaliter

notitia intuitiva non possit sine existentia rei, quae est vera causa efficiens notitia intuitivae mediata vel immediata.][35]

Ockham does not say here that, naturally, intuitive knowledge is *of* the existence of the thing, but rather that, naturally, one cannot have intuitive knowledge when the thing concerning which I have intuitive knowledge does not exist. That is to say, if I have intuitive knowledge of Socrates and whiteness, then I can know that Socrates is white. According to the order that God has created, that intuitive knowledge that Socrates is white cannot be had except when Socrates and whiteness exist. My intuitive knowledge is *not,* however, *of* existence, but rather existence is the necessary condition of my having intuitive knowledge at all.[36]

When Ockham turns to this question in his *Quodlibetal Questions,* there is a shift in terminology. There, he talks of presence and absence, not existence and nonexistence. He most certainly does not confuse presence and existence. Here presence and absence are seen to be what they are, namely, relations. He argues that God is not able to cause in us evident knowledge of a thing such that it would appear to us that the thing is present when it is absent.[37] Ockham's argument here is that if God could cause evident knowledge of a thing that is absent such that it would appear to us to be present, this would include a contradiction. The contradiction, however, does not arise from any insight into the nature of intuitive knowledge and God's absolute power. Rather, the contradiction arises because presence is a relation and a relation presupposes the relata. For God to cause a relation without the relata would be a manifest contradiction. For the relation is nothing other than the relata.[38]

Ockham's most telling interpretation of intuitive knowledge is that it is a cause of evident assent. "Evident assent denotes thus it is in reality just as is imported by the proposition to which there is assent" [Quia assensus evidens denotat sic esse in re sicut importatur per propositionem cui fit assensus].[39] Here, even though the example given is "this whiteness is" Ockham shies away from giving this a purely existential reading and prefers, instead, to read it as "thus it is in reality."

Existence for Ockham is the condition of intuitive knowledge. When viewed *secundum se et necessario,* it is about existential status. God could have created a world in which intuitive knowledge would be possible concerning a nonexistent object. In that case, I would be able to have intuitive knowledge of Socrates and whiteness such that I know that Socrates is not white—even if whiteness and Socrates do not exist or are not present. God, for example, has such intuitive knowledge, according to Ockham.[40]

For Ockham, then, existence is not an epistemological issue.[41] Ockham, therefore, does not have the problems relating essence and existence that Aquinas has. The entire theory of "esse," which is the celebrated accomplishment of Aquinas's thought, vanishes without a trace into Ockham's insistence that, because of God's absolutely freely chosen creative activity, existence is always a given and cannot be thematized philosophically. It can be thematized theologically, but then only in terms of creation. When creation is seen truly as creation, then existence ceases to be a problem. If Aquinas had a problem with "esse," it can only be the result of his attempt to hold the divine creative activity to some rational course. In the face of that, essence is a given (it is, in short, the presupposition of natural theology) and existence is a problem. For Aquinas, essence is both ontologically constitutive of things and it is also present in the divine idea that is the "exemplary cause" of the thing (hence the thing is the "middle" between our knowledge and God's). Therefore, *esse* must arise in his thought in order to distinguish the mode of being that essence has in an actually existing thing and in the divine mind.

One major problem in medieval thought can be seen as that of moving from a doctrine of creation that is essentially emanation to a doctrine in which creation is truly creation. We see here developments in two areas merging into one and the same field. On the one hand, as the theory of science moves away from a doctrine of necessity that would bind the activity of God and, indeed, of the world, science is allowed to range over contingent states of affairs. The pivotal point here is the move away from certainty based on necessity to certainty based on other means. On the other hand, the doctrine of God's absolute power, so crucial in the Condemnations of 1277, forced theologians to accept the absolute contingency of the world and to understand creation as creation. The result is that existence is the condition for the possibility of a science and existence, as creation, cannot be further thematized. The existence of things appears "without why" and, as a result, science based on such things can never hope to grasp their rational ground.

The Ockhamist Synthesis

It had been thought that the Thomistic synthesis brought together reason and revelation, philosophy and theology, Athens and Jerusalem. We have seen, however, that the synthesis never quite happened. Aquinas chose, instead, to lose Jerusalem to Athens. There can be no synthesis because one must always take priority over the other. The instability that revelation brings to a philosophical system cannot be tolerated. And it was not tolerated by Aquinas. Revelation made reasonable is natural theology—it is making God a part of the cosmos. This is the result of Aquinas's argument that theology is necessary.

After the Condemnations of 1277, thinkers were freed from the burden of attempting the synthesis—at least in the Thomistic method. The synthesis, however, was achieved along different lines. While Blumenberg is perhaps right that nominalism causes a certain despair in this world such that one must look outside of this world for salvation,[42] the despair is only in terms of finding God within the world. The human link to God is, for Aquinas, written in the cosmos and works in both directions—it links humans to God and God to creation. For Ockham, the human link to God can only come through faith since the only link we have is God's power to create *any* universe and not just this one.

Since, for Ockham, creation leads us nowhere but to the radical groundlessness of the divine will, we can approach nature with a new optimism, a new strength, and a new technique. Creation presents itself first as the ground of existence. Secondly, however, creation presents itself as gift to us. The gift, however, bears no connection to the giver—except in the tautological sense that a gift must be given. Therefore, nature as creation can be unpacked. And since we are cut off from God except in the act of faith, we might as well unpack that gift and make ourselves at home here.

In this sense we are no longer uneasy about the world, but we are encouraged to make our home in it. Thus we are placed on a new footing with regard to our knowledge of nature. This new approach is best carried out by experience and experiment. To be sure, one finds a role for experience in Aquinas's thought. He argues, along with Aristotle, that knowledge begins with sense. When I turn the object sensed into a phantasm and then abstract the universal from that phantasm, the universal cannot be the *product* of the object of sense. The universal must preexist the object of sense. Thus, the object of sense, the object of experience, is made sense of using those universals that come from elsewhere. For Ockham, the role of experience is completely different. Here, the universal *is* abstracted from the object of experience and does not preexist it. The result is that knowledge that, on the one hand, comes from experience, and, on the other hand, is universal (which is required of scientific knowledge), requires contact, often repeated, with nature.

The Ockhamist Solution

The solution Ockham offers to the problems that Aquinas bequeathed to subsequent thinkers is not without difficulties of its own. Indeed, Aquinas's thought leaves nature an ethical and almost divine status. Ockham's world and that of subsequent generations is devoid of any ethical status. It is into this world that science can penetrate, at times even violently. It is this world that becomes open to human destruction. And it is the epistemology of this world that leaves us with the modern epistemology where knowledge is production.

What must be approached with extreme concern, however, is the evaluation of the solution offered by Ockham. For that evaluation can come only either from the point of view of an earlier world, and that it is nothing other than fundamentalism, or from the point of view of a later world, and that is nothing other than narcissism. We cannot look at the decline of Scholasticism with either joy or sorrow. What is more, the problems that many would lay at the feet of Ockham are nothing more than the attempt to solve the problems of Aquinas himself.

The clash of the world of Aquinas and the world of Ockham is seen most clearly in the Reformation and its aftermath. In many ways, Luther is indeed a member of the school of Ockham. More significantly than that, he inhabits Ockham's world. The Catholic Church recognized this clearly. The response was not theological nor even philosophical—the response was to counter that world with another. The problem was, that world had already disappeared.

CHAPTER 7

AFTER OCKHAM:
MARSILIUS OF INGHEN AND
PIERRE D'AILLY ON KNOWLEDGE
AND THE EXISTING SINGULAR

After Ockham

The philosophy of the later fourteenth and early fifteenth centuries has been the focus of increasing attention and continued reassessment. Thinkers who were once labeled, usually derisively, "nominalists," are beginning to emerge as original, eclectic, and important thinkers in their own right. Historians and philosophers are beginning to realize that the divisions between the *via antiqua* and *via moderna* so common in the fifteenth and early sixteenth centuries cannot be applied accurately to the fourteenth century. Two philosophers to emerge as important in contemporary scholarship are Pierre d'Ailly and Marsilius of Inghen.

There is no doubt that Ockham's thought continued to exert influence—both positive and negative—for generations after his death. My choice to trace the influence of Ockham's thinking through Marsilius and d'Ailly, therefore, is in some ways arbitrary. There are two reasons why I have chosen these two, however. First, they are thinking at a distance of about a generation from Ockham.[1] Second, they were both trained at Paris and held influential positions there. They were responsible, to a large measure, for the extension of his views beyond the English Isle. Finally, their thought exhibits a reliance on that of Ockham, but they cannot be described accurately as "Ockhamists," for each criticizes Ockham on a number of issues and each presses his conclusions in directions not foreseen by Ockham himself. Together, they give us some idea of the medieval response to Ockham's philosophy of the existing singular.

Marsilius of Inghen

Marsilius was born around 1340, probably in the town of Nijmegen.[2] Marsilius began his career as a master of arts in Paris in 1362 and was a member of the English nation there. He began his studies in theology at Paris in 1366, though he did not complete them there. Marsilius held many administrative posts at Paris, including rector (1367 and 1371) and procurator of the English nation (1363, 1373 to 1375). In 1369 and again from 1377 to 1378, he was the representative of the University at the papal court in Avignon, a prestigious post. From 1379, Marsilius is no longer mentioned in relation to the University of Paris.

Marsilius became a master of arts at the University of Heidelberg—in fact, he was a founder of this university. He continued his career of service there, serving as rector nine times. In the 1390s, Marsilius resumed his study of theology, being the first to receive a doctorate of theology from that university. His reputation, already secured at Paris, continued to spread from Heidelberg. His works were highly influential in Erfurt, Leipzig, Prague, and Cracow. In several universities, it was through his writings that the *via moderna* was taught—perhaps thus securing his reputation as a nominalist in the derisive sense.[3]

Marsilius seems not to have completed his *Commentary* on the *Sentences* of Peter Lombard at the time of his death. He lectured on the *Sentences* from 1392 until 1394, but there is some evidence that he actually began working on it previously, maybe even in Paris.[4] The *Commentary* shows remarkable familiarity with thirteenth- and fourteenth-century theologians such as Aquinas, Henry of Ghent, Scotus, Aureoli, Ockham, as well as later fourteenth-century thinkers such as Wodeham, Gregory of Rimini, and philosophers such as Albert of Saxony and Jean Buridan.

Marsilius opens his prologue to his commentary with the typical question, "Whether theology is one science with God as its subject."[5] He divides the question into five parts: (1) the mode of generation of knowledge in us and the common divisions of knowledge, (2) what is theology, (3) whether theology is a science, (4) whether it is one science, and (5) whether God is its subject. In the first article he presents his theory of knowledge and treats the issue of *notitia intuitiva,* though he himself does not use this term in this setting.

Marsilius begins this investigation with a general definition or description of *notitia:* "Knowledge is the apprehension of a thing by a cognitive potency or it is an act or habit by which a cognitive potency knows formally by act or habit" [Notitia est apprehensio rei a potentia cognoscitiva. Vel sic: Est actus vel habitus, quo potentia cognoscitiva actu vel habitu formaliter cognoscit].[6] Marsilius then goes on to present a detailed and complicated set of divisions of various kinds of knowledge. The first general division is between knowledge that is the cause of things, and knowledge that is caused by the thing.[7] This is a division in kinds of knowledge that, as we saw, Aquinas also expressed. Yet for Marsilius, this division presents itself as more "absolute" than it seemed for Aquinas. For Marsilius, this division means that while God's knowledge is causative of things, our knowledge depends on things that are caused. But the dependency of our knowledge will result in its inability to ever be completely adequate to the things in the way that God's knowledge is.

This is clear as Marsilius continues his division of knowledge. For caused knowledge is divided into intellective and sensitive, that is, into those two manifest potencies of the soul. Sensitive knowledge is divided into simple and complex. Simple cognition is the apprehension of a singular thing. For example, external sense apprehends a colored thing by one simple apprehension of the singular, representing the thing with all its accidental properties perceptible by vision.[8] Notice that in this simple apprehension of the thing, all accidental properties are grasped in a single act. While at any given time I grasp all these properties, it is one act that "virtually includes" all of these that can be expressed as "this size," "this color," "this shape," etc. What is not grasped, according to Marsilius (following Ockham), is simply the color, size, shape, etc. Rather, simple sensitive cognition (and ultimately intellectual as well) grasps the thing that is colored, sized, shaped, and so on. In short, this is a knowledge of the individual, but a kind of knowledge in which it is grasped with all its accidental properties.[9] The accidents, then, are not grasped, except insofar as they are accidents of *this substance* or individual.

Complex sensitive cognition, on the other hand, is when sense says that this seen thing is this sweet thing. Marsilius refers this to the common sense, which can pertain at one time to many exterior senses.[10] As we shall see later, this kind of sensitive cognition will presuppose an act of simple, sensitive cognition.

Marsilius then provides a division of intellective knowledge that runs parallel to that of sensitive. Just as in sensitive, intellective knowledge can be either simple or complex. It differs from sensitive, however, in that simple, intellective knowledge can be either of the singular or of what is common.[11] Furthermore, it differs in that simple, intellective cognition can be of two kinds: vague and determinate. Vague, intellective knowledge follows sense, that is, it is knowledge of the individual as presented to sensitive cognition. While Marsilius does not further define this, we can understand that since its counterpart, determinate, intellective cognition will be of, for example, Socrates or Plato, this first kind can be called vague because what is given in sense is never Socrates or Plato determinately. Rather, sense, as we saw, presents the thing as colored, sized, and so on. Thus, the intellective cognition that follows on sense will be of the singular (though it can be of what is common) *as grasped through sense.* Since sense knows only these accidents as belonging to a subject, its intellective counterpart will know the individual only vaguely. Determinate, intellective cognition, on the other hand, is the most distinct of all simple knowledge. It corresponds to the individual in the category of substance. Common, simple, intellective cognition, finally, is the apprehension of a thing from its mode of signifying many things for which it supposits.[12]

What has been left unsaid to this point is the relation among these various forms of *notitiae.* Is there an order among them? Do some of them presuppose others? The order will be determined in two ways: time and perfection. That which is prior in time is more imperfect, that which is posterior more perfect. In this way, the sense of touch is temporally prior to the other senses because of its materiality and imperfection. Second, exterior sense is prior to interior, because interior sense presupposes exterior. Third, in intellective knowledge, vague knowledge of the singular is prior. For if we look at how the intellect knows through phantasms, these presuppose a knowledge that represents the thing just as it is in sense. Even among vague knowledge, however, there is an order of more and less distinct, with the less distinct being temporally prior to the more distinct. Fourth, among common, intellective knowledge, the more common is prior. For example, the concept of substance is prior to that of body. The highest would be the knowledge of the singular, which is most distinct. This kind of knowledge is also, according to Marsilius, the most difficult to attain. Just as for Scotus, the singularity of the singular is something that would exceed knowledge in that it would not be known in itself, here Marsilius holds that it is knowledge of the singular that falls in many ways outside the purview of this *notitia.* The more common and less distinct is always prior, and knowledge may never achieve a grasp of that which is in itself most distinct, that is, the singular.

Just what is the relation between "common knowledge" and knowledge of the singular? There are, according to Marsilius, no universals in being. Yet some singular things are essentially "more similar" than others. From this similarity, the intellect can make concepts "naturally and not freely."[13] That is, from these similar things themselves, the intellect forms a natural sign that is a concept. This concept arises according to the order of vague singular concepts according to their greater and lesser distinction. For example, from the concept of this extended thing, the concept of substance and body arises. For Marsilius, these universal concepts are nothing more than removing a certain kind of distinction from the concepts I form of singulars. These concepts of singulars, however, must be vague concepts, for otherwise the removal of distinction will not occur. I shall return to this point below, but first we must complete the divisions of knowledge.

The next division of complex knowledge is into that which is complex by "propositional complexity" and that which is complex by "nonpropositional complexity." The first would be the knowledge that is complex because it is knowledge of a proposition. The second is complex because it is knowledge of something that is of itself complex, for example, the soul, white man, and so on.[14]

There is then a division that pertains to knowledge of "propositional complexity." On the one hand, this knowledge could be apprehensive, by which the sense that the proposition imports by the signification of its terms is apprehended.[15] The fact that Marsilius designates this kind of complex knowledge "apprehensive," leads us to relate it to the definition of knowledge itself. For knowledge was defined as an apprehension of the thing. This kind of complex knowledge would grasp, as it were immediately, the sense of a proposition through the signification of its terms. There is no further act of the intellect necessary to this kind of grasping. It would be something akin to "intuitive knowledge" that has a proposition as its object.[16] On the other hand, this complex, propositional knowledge could be assentive, by which we assent to the proposition apprehended. For example, when a student, hearing the conclusion that the teacher wanted to prove, has apprehensive knowledge before the proof if he understands what it signifies, that is, the student assents while he does not have proof.[17]

Assentive knowledge, then, is divided based on whether there is prior proof of the proposition or not. Assent without proof comes in four ways. First, there is assent without proof to singulars, which is called sensory, assentive knowledge because it follows upon the senses. For example, when the thing touched is hot. Second, the assent may follow experience with the aid of the intellect, for example, every fire is hot. Third, the assent can arise from the evidence of principles from the implication of their terms, for example, the whole is greater than its parts. Fourth, assent can be generated from the authority of the one speaking. Marsilius now gives names to these categories of assentive knowledge. The first is called "sense" or "following sense," the second and third "*intellectus,*" and the fourth "faith."[18]

The final division,[19] assentive knowledge that arises with proof, follows the nature of the middle term by which it is proven. This also is divided into four categories. One kind proceeds from signs somehow moving the intellect, but the intellect is not moved firmly. This is called suspicion. Another kind proceeds from what is probable into a consequence that is necessary by other inferences by which the intellect gives assent to the consequence, but fear remains concerning the consequence because the premises were only probable and not necessary with respect to the consequence. This is called opinion. Another kind is from primary and necessary premises, either immediate or mediate with respect to the conclusion. This is called science, which is from necessary truths and concluding about necessary truths. Finally is faith, which comes from premises that are believed and concludes with that which is necessary.

This last categorization poses some difficulties. For unlike his previous interruption into the division of knowledge, Marsilius does not pause here to talk of the relation that obtains between these various kinds of knowledge. For instance, from where does the knowledge of the primary and necessary premises of scientific knowledge arise? A step was made from nonpropositional, intellective knowledge to propositional in an earlier division. Thus, these later divisions are all divisions of propositional knowledge. But do these kinds of *notitia* presuppose others? And if so, which kinds?

If we trace the line from scientific knowledge (i.e., intellective, propositional, assentive knowledge that comes through previous proof) backwards, we can begin to see the outlines of these relations. Why is it that we assent to the scientific proposition, that is, the conclusion of the scientific demonstration? Marsilius tells us that it is because this proposition arises from primary and necessary premises. But how do we come to acquire knowledge of these primary and necessary premises? The assent[20] must come from assentive knowledge that is not proven. For if it does not, then we enter an infinite regress of demonstrations—the precise situation that Aristotle's theory of *scientia* was designed to avoid. Indeed, Marsilius has one division of assentive knowledge that is knowledge without previous proof for the "evidence of principles." This division arises only from the "implication of the terms" of the principle.

Continuing our quest for the evidence of assent, we are forced to turn from this last kind of knowledge to propositionally complex, intellective knowledge that is knowledge of the sense of a proposition that is imported by the signification of its terms. The fact that Marsilius relies here on the common definition of *per se nota* propositions indicates his understanding of this as a necessary stage on the way toward science. But how is it that I could grasp the signification of terms? We seem thrown back to simple, intellective knowledge.[21]

Once we see that this complex, propositional knowledge must arise from simple, intellective knowledge, we can follow Marsilius's own path of the relation among these various forms of knowledge. This intellective knowledge of singulars can also be twofold: of singulars or of something more common. Yet we have already seen that the intellective knowledge of what is more common depends on the intellective knowledge of singulars and of their similarity. Ultimately, we are thrown back to intellective knowledge of singulars. Marsilius has already shown that this knowledge is probably not "determinate" for this is the most difficult kind of simple knowledge to attain. Furthermore, determinate knowledge of a singular is not required for a science. If vague knowledge immediately represents sense, and sense is an apprehension of the thing in which all its properties are "virtually included," then determinate knowledge could only give us knowledge of the singularity of the thing. Such a knowledge could not, by definition, lead to scientific knowledge of what is common to many singulars. Determinate knowledge, therefore, gives knowledge of the singular, but science has no concern with the singular, it is concerned only with the conceptual content of our grasp of things. Only this grasp can lead to the universalizing knowledge that is *scientia*.

Marsilius, much like Scotus and Ockham, points to the singularity of the singular as an element that resists the grasp of knowledge. The intellective knowledge, which receives what is sensed, is vague knowledge. If science is ultimately built up from these blocks, science will never be of singulars. Singulars as singulars would stand outside the domain of the conceptual structure of scientific knowledge. In this way, science would not uncover the rational ground of singulars because there is always an aspect of the singular—its very singularity—that would stand outside this kind of knowledge.

Marsilius, therefore, follows the lead of Scotus and Ockham. He, too, moves away from science as the knowledge that uncovers the causes of things and toward science as certain knowledge of the conceptual content of things—of unpacking what is "virtually included" in our grasp of individual things. As long as we recall, however, that science leaves aside the determinate knowledge of the singular, we also realize that the conceptual framework of science leaves aside the totality of what it means to be a singular. The conceptual unpacking that Scotus, Ockham, and Marsilius call *scientia,* must always be mindful that there is something about things in the world that resists that very conceptual unpacking. Concepts, and science built on them, are never adequate to their objects.

Pierre d'Ailly

Pierre d'Ailly was a generation removed from Ockham himself. Born in 1350 or 1351 at Compiègne, he entered the College of Navarre at Paris in 1363 or 1364.[22] In 1367, d'Ailly received his licentiate in arts. The following year he began his studies in theology while continuing to teach in the arts faculty. He lectured on the Bible in 1374 to 1376 and on the *Sentences* in 1376 to 1377. He became doctor of theology in 1381.

D'Ailly was also quite involved in the administration of the university: He served as proctor of the French nation, carried the *rotolus* of the French nation to Pope Clement VII in Avignon, and was rector of the College of Navarre (1384) and chancellor of the University (1389). He was also involved in the political activities of the Church, especially the Schism, attempting to heal this obvious wound in the Church. He was named cardinal in 1411 and participated in the Council of Constance, which finally put an end to the Schism. He died on August 9, 1420.

D'Ailly's *Commentary* on the *Sentences* was delivered in 1376 and 1377. Paris was already under the influence of Ockham by this time. His commentary shows the heavy influence of Ockham, though d'Ailly is not shy about straying from Ockham's thought. His work also mentions Gregory of Rimini, Adam Wodeham, Robert Holcot, and other later fourteenth-century authors. His commentary was quite popular, appearing in seven editions by the end of the fifteenth century.

D'Ailly's prologue does not follow Ockham's exactly. He does, however, take up many of the issues that were most important in Ockham's prologue.[23] His first question has three articles: (1) whether it is possible that the viator have evident knowledge of some truth, (2) whether it is possible that the viator have knowledge greater than faith of theological truths, (3) whether it is possible that the viator have scientific knowledge of theological conclusions. These three questions force d'Ailly to engage in discussion about epistemological issues, and in this too he follows Ockham's lead. However, his commentary shows a development of Ockham's thought in new directions, or at the very least a deepening of some of the issues. The most important, for d'Ailly, is the concept of evidence.

We have seen that Scotus introduced the concept of evidence to speak of the relation between simple knowledge and more complex conceptual structures (propositions, syllogisms, science) built out of that knowledge. Ockham used this concept to discuss the way in which propositions could be known with certitude. D'Ailly expands the discussion of this topic, giving it more depth and expanding its role in the theory of knowledge.

He begins his explication by immediately dividing evidence into two kinds: absolute and conditioned (which he also terms *secundum quid*).[24] Absolute evidence is defined as "true assent, without fear, naturally caused, by which it is not possible that the intellect assent and in thus assenting be deceived or err" [Evidentia absoluta simpliciter potest describi quod est assensus verus sine formidine causatus naturaliter quo non est possible intellectum assentire et in sic assentiendo decipi vel errare].[25] This definition has many parts that specifically define evidence against other kinds of assent. For all evidence is assent, but not all assent is evidence.[26] Only evidence is true assent. Second, this assent is without fear or compulsion, which distinguishes evidence from opinion and conjecture. Third, this assent must be naturally caused, that is, the cause of this assent necessitates the assent. This distinguishes the assent from faith, which is not caused naturally but freely. Finally, the definition includes the fact that one cannot have this assent and doubt or be deceived in assenting. This distinguishes evidence from the assent I might give to a proposition that is the conclusion of a falsely written proposition. In this way, by assenting to a proposition that is false but forms part of a valid syllogism, I might assent to the conclusion and be deceived. This last part of the definition also distinguishes this kind of evidence from conditioned evidence.[27]

If this last condition distinguishes conditioned evidence from absolute, then doubt might be possible in conditioned evidence. This possibility arises in conditioned evidence precisely because of the "condition" that it entails. This kind of evidence is subject to a form of doubt because it only holds given the "standing, general influence of God and no miracle being done" [stante dei influentia generali et nullo facto miraculo].[28] D'Ailly refers to this kind of evidence as "created." It is the kind of evidence we have of everything that is exterior and contingent.

There are, according to d'Ailly, many possibilities for the viator to have absolute evidence. First and foremost, absolute evidence is had of the principle of noncontradiction (the "first principle"). But above and beyond that, d'Ailly argues that such evidence is possible of other things. For example, one can know that one is, that one knows, and other similar things by absolute evidence. But of something extrinsic, knowledge of which comes through the senses, absolute evidence is impossible.[29] This kind of knowledge is always subject to doubt. For there is no contradiction in saying that "whiteness is" appears to someone and that person assents that it is thus (whiteness is) and yet it is not thus in reality (whiteness is not). D'Ailly raises here an argument that is similar to that Ockham raised to argue for the possibility of intuitive knowledge of a nonexistent thing: God could destroy the thing sensed but maintain the sensing. Statements such as this are precisely what leads commentators to charge later medieval thought with skepticism. Yet notice that d'Ailly does not take the easier skeptical route, that is, he does not state simply that our senses can go wrong. Rather, he supposes that knowledge is naturally inserted within a world in which the soul reaches out to its object, and similarly with sense. Therefore, in order to produce error, the absolute power of God has to be introduced. This kind of error, however, must presuppose the suspension of the "general influence of God," or must presuppose something miraculous. This kind of error does not happen naturally, but conditioned evidence does happen naturally.

Rather than being productive of skepticism, this introduction of a kind of doubt arising from divine action presupposes the functioning of knowledge without doubt. It takes some action on God's part to introduce this doubt. Naturally, I cannot doubt that of which I have conditional evidence. The fact that all of my knowledge of contingent, external facts is conditioned on the "standing, general influence of God" and the absence of miracles highlights two issues: First, our knowledge of such things cannot ever get behind the things to their causes, and second the contingency of the world means that exterior, contingent things are always referred to the divine will.

D'Ailly further illustrates this point by arguing that as long as we do not take into account the miraculous, that is, as long as nature follows its course, conditioned evidence is of such a nature that, when we have it, we cannot rationally doubt.[30] If we take into consideration the general influence of God, then nothing can appear to us and not be as it appears. For this is the very course of nature God has instituted.

D'Ailly later fleshes out some of the results of this division of evidence. He asserts that only God's knowledge is infallible. Yet, he immediately states that some human knowledge is similarly infallible. Knowledge of the principle of noncontradiction or other things that are necessary and unable to be otherwise is of such a kind. For in these cases, it is not possible to have such knowledge and yet the state of affairs be otherwise than is signified. But to speak of a "state of affairs" here is misleading. For obviously such truths (as the principle of noncontradiction) do not refer to states of affairs. Rather, they refer to the very possibility of truth itself. Without the principle of noncontradiction, there can be no truth, and, consequently, no assent. The fact that one may dissent from the first principle does not mean that the first principle is not evident, but only that it can be not evident. One who has evidence of the first principle would not dissent. D'Ailly sharply distinguishes between the *possibility* of dissent to such truths and the *act* of dissenting to such truths.[31] There are other contingent truths, as we have seen, that can be known infallibly, for example, that I live, that I know, that I am, and so on.[32]

If d'Ailly argues that conditional evidence of exterior contingent truths is all we can hope for in this life, and furthermore, if conditional evidence does not leave room for "rational doubt," then what is the purpose of the distinction at all? It is used to call attention to the difference between divine knowledge and our knowledge. Since God's knowledge is always infallible and ours is, in many cases, fallible, the distinction in kinds of evidence points to the fact that our knowledge of exterior, contingent truths is always "running after" the truths known. Our knowledge, the distinction tells us, depends on an act of the divine will—an act that is never open to knowledge itself. Our knowledge of contingents, therefore, has a condition that is, by definition, unknowable—we can never know whether things stand under the general influence of God or not.

D'Ailly's distinction between these two categories of evidence points to a more general concern with the way in which the power and will of God are related to existing things. If the will of God is not subsequent to the intellect of God, if God's act of creation is best grasped as an act of the will, then God's creative activity itself will be "without why." This creative activity will leave its trace in the things that are created. The trace of the divine will, of the power of God, is precisely the fact that our knowledge of things is *conditioned*. If created things appear "without why," then we can no longer rely on the rational ground—the metaphysico-epistemological notion of causation—to ground our knowledge. Our knowledge of things (as opposed to necessary truths) is fallible because the world of existing things is always contingent—it always bears the trace of the divine will.

Conclusion

To be sure, the Condemnations of 1277 stand as a watershed in medieval thought. The emphasis on divine power that is ushered in by this theological reaction to philosophical excess, however, had deep and far-reaching philosophical results. The condemnations unwittingly pointed out the failure that philosophy, as the search for the rational ground of existing singulars, will always suffer. For the rational ground is a demand that *we* place on singulars—it does not belong to their own mode of being. God's power can be used to signify this fact. Once existing singulars are referred to the divine will rather than the divine intellect, the failure of reason to grasp their ground becomes evident.

The search for the rational ground, therefore, founders when singulars qua singulars, are offered up against the ground discovered by reason. Yet the thinkers who follow in the wake of Aquinas also show an unwillingness to offer what might be called a "philosophy of the singular." Instead, we have seen Scotus, Ockham, Marsilius of Inghen, and d'Ailly simply point to the various ways in which the singular remains untouched in our conceptual and scientific grasp of it. Our knowledge, they argue in various ways, is not adequate to singularity. This realization, however, has two consequences. First, there can be no concept of a singular or singularity. The attempt to thematize singularity directly would be destructive of that very singularity itself. Second, our knowledge still operates with things in the world, but its mode of operation always "follows after" the things and never functions as the ground of their being.

We have chosen an arbitrary end point for this analysis. Certainly the question could be pursued as universities become the site of the contest between the *via antiqua* and the *via moderna*. The question could be pursued as Thomistic thought is reestablished in order to answer to Luther. The question could even be pursued as the dawn of modern philosophy takes up the task of regrounding existing things in the knowing subject (the "cogito"). Finally, the question could be pursued into our own century, as philosophers have discovered, as if for the first time, that the search for the rational ground of existing singulars is doomed to failure. Yet as we pursue the question into all these areas, the insights of late medieval philosophy can serve as a model and a warning—objects do not go into their concepts without remainder, but without concepts, power appears as its own reason.

UNCONCLUDING POSTLUDE

It has never been immediately obvious to me, which is to say that it still stands as a philosophical question for me, why we should continue to read philosophical texts from the Middle Ages. For some, this question is easily answered by pointing out that these thinkers, or at least one of them, was "right." The ease of this answer has always puzzled me, for it assumes that history itself does not make any difference in our judgments about philosophical positions and their "rightness." I am convinced that even if rectitude is a proper evaluative tool for the historian of philosophy, rectitude itself must be judged on the basis of history, and not on the basis of an a- or extra-historical concept of reason. The texts of the past are like intentionless stars that we light up by reading, as one might say following Benjamin and Adorno.

Thus we began with a lie. The lie was that we would be looking at the historical field in reverse, that is, from the point of view of Ockham and his followers. For in fact, we have been looking at the field from the point of view of the question, posed most forcefully by Heidegger and Adorno, of the relation between the philosophical pursuit of a rational ground and the given existence of singular things. The question itself, that is, the posing of a question at the very heart of the positing of a rational ground, arises with great seriousness only in the twentieth century. It is only with the possibility of posing the question of rational ground that this study can get underway, since the question of rational ground as such was not directly addressed in medieval thought.

This opens a serious danger that, in reading the texts of medieval thinkers through the lens of a contemporary question, the texts will be forced to speak a language that is foreign to them. Yet this is the unavoidable situation of engaging in a reading of the history of philosophy, if history itself is taken seriously. If history itself is taken seriously, then we are already aware that our reading of medieval texts comes from a contemporary historical site. The danger, therefore, lies in reading medieval philosophy without reflecting that reading through the lens of history.

The calling into question of the rational grounding of existing singulars arises in modernity. The fact that the questioning of ground arises in modernity, however, does not entail the fact that the *activity* of grounding is also peculiarly modern. Only once the question of the relation of rational ground to existing singular is posed, can we then return to the history of philosophy such that the grounding of existing singulars in reason can be uncovered. This return, reflected through history, does not require a second step of bringing the medieval texts back to speak to us, for in the return they are already speaking to us. I would here like to listen, for a moment, to what these texts have said to our question, the question of the grounding of existing particulars. In order to do so, I first need to pose the question with somewhat more specificity.

The question of rational ground, that is, the question of the identification of the being of things with our knowing things, is posed with the greatest clarity in the age of the primacy of the subject. For when a knowing subject is posited as standing securely over-against objects whose being is in need of proof, then the question of the relation between the subject and the object comes immediately to the fore. The move to subjectivity, as found, for example, in Descartes's *cogito,* is a move to show the security of the being of the subject in its knowing prior to the being of things outside the subject that are known by the subject. From Descartes follows an entire tradition of attempting to prove the existence of the objective world precisely because that world was left behind in the attempt to secure the being of the subject. Once subjectivity is posited and secured in advance of objectivity, it is clear that the being of the object will be identical with the mode of knowing of the subject. For this subject was posited precisely as knower, and secured in its being as knower. Therefore, the only route possible for a return to things is through knowledge. Modern thought from Descartes through Hegel and up to Husserl exhibits the identification of being and knowing, that is, it openly posits the identification of being and knowing.

The fact that modernity exposes this identification primarily in terms of subject and object does not mean that a similar identification of being and knowing was not present in philosophy before—albeit in a different form. For Heidegger, the move in modernity is but the culmination of the tradition of philosophy that stretches from Plato to the twentieth century, a tradition called simply "metaphysics."[1] Even though Leibniz was the first to formulate the principle of ground, that is, the principle of sufficient reason, his formulation merely brings to expression the principle upon which metaphysics itself is situated: Nothing is without reason.[2] According to Heidegger, this principle merely gives expression to the fundamental identification of knowing with being that characterizes metaphysics from its outset in Greece through its intensification in modernity.

I have argued that Aristotle's theory of *episteme* is built on the unexpressed principle of ground to the extent that it pursues the principles of being along the lines of the principles of knowing. The dual role that form (and indeed essence) plays in Aristotle's thought—that it is responsible for the being of a thing but is also that which is known when the thing is known—indicates the presence of ground in his thought. Yet if Aristotle's thought is based on the positing of a rational ground of existing singulars, why is the principle of ground left unstated by him? Precisely because Aristotle's thought does not operate on the basis of a distinction between subject and object. Since the identification is implicit, it cannot be brought to explicit expression.

If substantial form is the expression of rational ground (an expression that has no need for the principle of sufficient reason) then the moment when substantial form comes under attack should be a crucial moment in the history of philosophy. It is this moment that I have charted here through the question of whether theology is a science. Tracing the question of ground in relation to substantial form by way of the question of the scientific status of theology allows for the exposure of the question of ground precisely where it is most crucial in medieval Christian thought: the site where God is thought as creator of the universe.

Substantial form can function as rational ground because of its posited connection to universality. Since rational knowledge has a universal as its proper object, substantial form as a principle of being is also posited as universal.[3] Yet when substantial form is posited *both* as universal *and* as a principle of being, then knowledge of substantial forms must be pushed back into the divine mind as a pattern for creation. In this way, God is the name for the rational ground of existing singulars *insofar as the divine intellect is given priority over the divine will*. It is now clear that in this mode of providing a rational ground, there is no need to secure the subject before returning to secure the being of objects. For God as creative intellect secures the identity of knowing and being in that very creative activity.

Ockham's nominalism (and here Scotus is his precursor) attacks the rational ground of substantial form from two directions at the same time. On the one hand, Ockham rejects the notion that God's intellect is prior to the will, even if only conceptually prior. When God's will is brought to the fore in the analysis of creation, then God's knowledge cannot function as a principle of being. Instead, God's will is the sole principle of being, and therefore in the act of creation the principles of knowing and being are not identical. On the other hand, Ockham also attacks the notion of a universal substantial form existing outside the soul. For him, the universal is a result of the soul's activity in relation to the givenness of singulars in the world. Consequently, while a universal form may be the principle of knowing a thing, it is not at the same time the principle of the thing's being.

To this point, Heidegger's insistence that rational ground is always *given* to things and is, as such, not itself grounded has led the reading of medieval thought presented here. Yet while Heidegger's recognition of the *ab-grundlich* character of ground has opened this issue for my reading, it ultimately has to be left behind if we are to make sense of Ockham's nominalism. Yet more than this, it must also be left behind if we are to accept Ockham's position that things outside the soul are immediately and irreducibly singular. For while Ockham recognizes—particularly in his notion of intuitive knowledge—that existing singulars are not completely and adequately grasped by our conceptual knowledge of them, he still insists that there is a relation between knowledge and the object of knowledge such that knowledge can be called "evident." While concepts are anterior to the being of things, they still have a relation to the things of which they are the concepts.

The concern that has been raised over Ockham's alleged skepticism arises precisely because Ockham denies, on the one hand, that what is grasped in a concept (particularly a universal concept) is constitutive of the being of the thing grasped and also, on the other, does not make the move into the subjectivism of a thinker like Descartes. From the point of view of Heidegger's thinking through the notion of rational ground, Ockham would seem to be a thinker who recognizes the groundlessness of our rational grasp of existing singulars. That is, the ground of rationality, from Heidegger's point of view, is not itself rational. Ockham, however, refuses to take this route—that is, he refuses to posit the fundamental irrationality of reason.

In this way, a second lie that stands as the basis of this investigation is exposed. At the outset I posed the question of rational ground as if Heidegger and Adorno stand in agreement on this issue. Indeed, they do agree to the extent that both recognize the nonidentity of the principle of knowing (the "concept" in Adorno's language) with the principle of being (the "object" in Adorno's language). Yet they part company in the most fundamental way when it comes to the philosophical consequence of that recognition. The reading of medieval reflections on the scientific status of theology and the role of the singular in science opened up by Heidegger's notion of the ungrounded character of ground and Adorno's notion that "objects do not go into their concepts without remainder" now can be turned back to this contestation between Heidegger and Adorno. While I am not able to fully address this contestation, I wish to point to some of the groundwork that this reading of medieval philosophy lays for thinking it through.

It is almost commonplace today to speak of the death of the subject, that is, to speak of the philosophical dismantling of the epistemic subject (first seen in its foundational role in Descartes's cogito). It is by way of this subject that philosophy in modernity attempts to bring about the identity of knowing and being, that is, to uncover the rational ground. From the point of view of dismantling this subject, this last attempt at rational ground, being can only appear as singular, as always characterized by difference, and as appearing to us without a ground, without a why. What seems possible—as if for the first time—at the moment of the death of subjectivity is to recognize that existing things exceed the rational concepts we have of them. And it is all too tempting to take this recognition and turn toward a philosophical exploration of that which exceeds our rational grasp.[4] Thus we find a major trend of philosophizing after the death of subjectivity that still follows the two poles of being and knowing, of object and subject. On the side of knowing, thinkers like Heidegger and Derrida expose what Derrida would call the "logocentrism" of Western philosophy from its inception. On the side of being, they will expose the way in which it is groundless, that is, that it appears without why or appears only as differing and a trace of deferred presence.

By orienting my reading of medieval thought—and Ockham in particular—with a reading of *both* Heidegger *and* Adorno, I was able to uncover in Ockham a different strategy of thinking the relation of knowing and being. For Adorno's strategy for thinking on the basis of the recognized inadequacy of rational concepts to get at the ground of being is not to posit a new mode of being that is now adequate to our concepts, but rather to trace the ways in which objects, existing things, will always have a remainder that cannot be thought through rational conceptualization.[5] Adorno's concern with Heidegger (and mutatis mutandis with Derrida) is that notions like *Ereignis* and *differance* do little more than reinscribe the identity of knowing and being. For what Heidegger actually does is to posit being as precisely that which is adequately mapped and thought because thought is no longer able to be the thinking of the rational ground of things. In other words, Adorno would argue that something like the eventing of being or *differance* comes to the fore only on the basis of our rational conceptualization of existing things. To now reorient thought toward this excess is to repeat the very way of thinking that defines metaphysics. In short, if Heidegger defines what he calls metaphysics as the discovering of the rational ground without recognizing that it is posited, then to rethink being according to the way in which we think is once again positing that ground, even if we no longer conceive thinking as rational and therefore must think being as event.

It is at this juncture that Ockham's theory of science, his use of God's absolute power as a methodological tool, and his theory of concepts (including universal concepts) appears interesting for our contemporary issues. For Ockham's insistence that creation is an act of the will primarily, and therefore not open to rational inspection, means that existing things have to be understood as simply present to the soul engaged in knowing. Second, his theory of knowledge and concept formation (which ultimately are one and the same for Ockham), presuppose this givenness and attempts to secure knowledge of the thing, *but not of its mode of givenness.* Yet for Ockham and those following in his footsteps, there is never a temptation to give up rational conceptualization in favor of thematizing givenness as such. Rather, his theory of knowledge and his theory of science always remain on this side of the existence of singular things. In this way, Ockham and Ockhamism hold open a strategy for thinking that relies neither on the rational ground of substantial forms, nor on the ground of the epistemic subject, and therefore never runs the risk of recapturing the mode of existing of objects that can only be seen by us as given. There is indeed for Ockham a remainder that stands outside our concepts, and the only way to be attentive to and redeem that remainder is to let it remain.

NOTES

Introduction

1. Plato, *Phaedo,* translated by David Gallop (New York: Oxford University Press, 1993), p. 96a9.
2. I borrow the idea of rational ground from Martin Heidegger's reading of Leibniz. The idea appears in, Martin Heidegger, *Metaphysische Anfangsgründe der Logik in Ausgang von Leibniz,* Gesamtausgabe, vol. 26 (Frankfurt: Vittorio Klostermann, 1978) esp. pp. 135–284 (translated as Martin Heidegger, *The Metaphysical Foundations of Logic,* translated by Michael Heim [Bloomington: Indiana University Press, 1984]) as well as in Martin Heidegger, *Der Satz Vom Grund* (Frankfurt:Vittorio Klostermann, 1997) (translated as Martin Heidegger, *Principle of Reason,* translated by Reginald Lilly [Bloomington: Indiana University Press, 1991]). However, there are many thinkers in the twentieth century who have addressed the issue of what I call here the rational ground. Theodor Adorno, e.g., raises a similar issue in his *Negative Dialectics* when he says that "objects do not go into their concepts without remainder." Theodor Adorno, *Negative Dialektik,* edited by Rolf Tiedemann, Gesammelte Schriften, 20 vols. (Frankfurt: Suhrkamp Verlag, 1997), 6: 16–17 (translated in Theodor Adorno, *Negative Dialectics,* translated by E. B. Ashton [New York: Continuum, 1992], p. 5). The issue that both Heidegger and Adorno raise is one of the relation between objects (or existing singulars) and that which makes the objects intelligible. When their principle of intelligibility *is* the reason or cause for their existence, then that principle is a rational ground.
3. The translator often translates the Greek "*aitia*" as "reason," though sometimes he uses "cause." The point is, however, that in the end these are one and the same.
4. In this context, one can compare the cosmology of Aristotle's *De Caelo,* in which there is no "prime mover," with that of the *Metaphysics.* The *De Caelo* replaces *nous* as prime mover with the concept of natural place in order to show why things come to be in the way that they do. Aristotle ultimately came to see that natural place cannot perform this function because it, in turn, needs a rational ground. His solution, much like the one

offered by Socrates in the *Phaedo,* is that *reason itself* is the reason why all things come to be, exist, and pass away.

5. It is in this way that Plato, Aristotle, Aquinas, and Scotus (just to name a few examples) must all search for a principle of individuation. What is strange is that the rational ground is supposed to be the explanation for the singular and yet once it is posited it is found to be in need of a principle of individuation. This is a sign that the rational ground cannot, on its own, account for the individual of which it is supposed to be the ground.

6. Often the ground that competes most directly with the rational ground is that of power or force. This is precisely the contest that happens whenever one wants to use philosophy to help explain God's creative activity. For in creation, the ground will always have an element of sheer power that is not subject to further understanding in terms of its "reasons."

7. In this relation, it would be interesting to reinsert the passage of the *Phaedo* into the dialogue as a whole. Socrates ends his discussion precisely with a turn toward myth. The end of the dialogue, therefore, recasts this passage in an ironic light. Viewed in this way, this passage of the *Phaedo* raises not only the search for rational ground, but at the same time erases this search as inescapably doomed to failure.

8. In Hesiod's account in *Theogony,* the reason most often given for the coming to be of something is *polemos*—strife or conflict. His account is one in which the power of the gods is not subjected to further questioning. The answer to the question *is* divine power and it seems as if this answer, in singular, does not appeal to reason.

9. Heidegger draws this phrase from a poem of Angelus Silesius, *Der Satz vom Grund,* p. 61.

10. This recognition that the ground is without ground or provides only a groundless ground, is referred to by Heidegger as *Ab-Grund.* This notion receives perhaps its fullest treatment in Martin Heidegger, *Beiträge Zur Philosophie (Vom Ereignis),* edited by Friedrich-Wilhelm von Hermann, Gesamtausgabe, 65 (Frankfurt: Vittorio Klostermann, 1989).

Chapter 1

1. Aristotle, *Analytica Posteriora,* edited by Laurentius Minio-Paluello and Bernard G. Dod, Aristoteles Latinus, vol. IV (Brussels: Desclée de Brouwer, 1968), p. xi.

2. Aristotle, *Posterior Analytics,* translated by Jonathan Barnes (Oxford: Clarendon Press, 1975), p. x. Hereafter Barnes translation will be cited as *APo,* followed by page and line number. His commentary will be cited as *Posterior Analytics,* followed by page number.

3. Jonathan Barnes, "Aristotle's Theory of Demonstration," in *Science,* Articles on Aristotle, vol. 1, Jonathan Barnes, Malcolm Schofield, and Richard Sorabji, ed. (London: Duckworth, 1975), p. 65.

4. Barnes has argued that the "the doctrine expounded in the *Posterior Analytics* is entirely—or almost entirely—independent of the Syllogism," Jonathan Barnes, "Proof and the Syllogism," in *Aristotle on Science: The Posterior Analytics*, edited by Enrico Berti (Padua: Editrice Antenore, 1981), p. 8. His view is that while certain passages of *APo* presuppose knowledge of syllogistics, it makes sense to see these as later additions (see especially p. 28). More recently, Robin Smith, "Immediate Propositions and Aristotle's Proof Theory," *Ancient Philosophy* 6 (1986): 47–68, has offered a different and deeper connection between *Prior* and *Posterior* Analytics. He wants to show, in fact, that Aristotle's concern with scientific demonstration is what led to certain developments in *Prior Analytics*. In the Middle Ages, however, it was assumed that *Prior Analytics* was thematically *and* logicially prior to *Posterior Analytics*.

5. *APo* 79a16–24. Smith ("The Syllogism in *Posterior Analytics* I," *Archv für Geschichte der Philosophie* 64 [1982]: 113–35) has argued that this passage refers not directly to the *Prior Analytics*, but to a simple form of the syllogism. Barnes makes a similar point in "Proof and the Syllogism," 34–57, where he argues that the main theory of *Posterior Analytics* was developed prior to the syllogistic theory of *Prior Analytics*.

6. *APo* 71b20–25.

7. *APo* 73a25ff.

8. Aristotle, *Posterior Analytics*, p. 94.

9. Aristotle, in fact, replaces "immediate" with "indemonstrable" (*anapodeiktos*) when he goes on to explain this requirement at 71b27.

10. Aristotle, *Posterior Analytics*, p. 100.

11. Ibid.

12. James A. Lesher, "The Meaning of *nous* in the *Posterior Analytics*," *Phronesis* 18 (1973): 44–68, points out that for Aristotle *nous* has many tasks to perform and only one of those tasks is to grasp first principles.

13. This translation has been slightly modified, replacing Barnes's "skill" with "art" and his "understanding" with "science." This is not to quibble over Barnes's choice of terms, but rather is to make the text uniform with medieval translations of and commentaries on the text.

14. Richard D. McKirahan, *Principles and Proofs: Aristotle's Theory of Demonstrative Science* (Princeton: Princeton University Press, 1992), p. 239.

15. McKirahan, *Principles and Proofs: Aristotle's Theory of Demonstrative Science*, p. 256.

16. The beatific vision is not itself propositional in character, but might result in the formation of true propositions in the intellects of the blessed. Aquinas does argue, as we shall see, that only humans reason through discourse, and that does not rule out the possibility that humans enjoying the beatific vision also could reason through discourse.

17. This position begins with Boethius and is only augmented by Aquinas's assumption of Averroes's argument about the distinction between essence and existence. For all the attention paid to Aquinas's theory of

esse, medieval concerns about existence are widespread. As we shall see, it is precisely the relation of essence, which is in many ways the object of the rationality of *episteme,* to existence, which always seems to fall outside the grasp of reason, that is the central question here.

Chapter 2

1. See note 1 of chapter 1.
2. Robertus Grosseteste, *Commentarius in Posteriorum Analyticorum Libros,* edited by Pietro Rossi (Firenze: Leo S. Olschki, 1981), pp. 111ff. In the following, I will refer to this text by page number alone.
3. I do not mean to imply that non-caused and self-caused are interchangeable—for they are radically different. Aquinas, e.g., demonstrates in his first proof for the existence of God that there exists a mover that is not in motion itself—i.e., God is not caused. Spinoza, on the other hand, assumes that God is "causa sui." However one understands the first cause, the first cause would have being *per se.*
4. The term Grosseteste uses quite frequently is *"egredior,"* which has several senses. Grosseteste means by this term the fact that the essence of one comes out of the essence of the other. But *"egredior"* seems to be used to express a special connection between the two. For example, in a nautical sense, *"egredior"* means to disembark from a ship. Again, in a military sense, it means to march out of a city, a fort, etc. What this term is meant to express, and this will be made clear below, is the fact that the essence of the one is contained in the essence of the other and in that sense it "climbs out" of that first essence. Therefore, this term is meant to express precisely this connection of not only being contained in, but also coming out of, as from vessel or shell. Perhaps the best metaphor for this connection would be the relationship between a turtle or a snail and its shell. The turtle or snail may be pulled out of the shell, but it is contained in the shell as in its home.
5. The text reads: "Illud autem cuius quidditas essentialiter et non per accidens a quidditate alterius egreditur esse suum habet ab eo a quo egreditur sicut a causa vel efficiente vel formali vel materiali vel finali" (111).
6. Ibid.
7. The entire passage reads: "In hanc primum modum essendi vel dicendi per se alterum de altero cadunt omnes ille predicationes quae oblique predicant per se causam vel efficientem vel materialem vel formalem vel finalem de suo causato. In omnibus enim istis modis predicandi est predicatum tale quod ipsum recipitur in diffinitione subiecti quod ab ipso predicato habet esse."
8. This is not necessarily true of all definitions, but only of real definitions. The case is somewhat more complicated with nominal definitions.
9. Grosseteste's commentary is divided into *"conclusiones,"* which draw out the main points of Aristotle's text. The sixth conclusion states: "Omnis demonstratio est syllogismus ex necessariis, omnia et sola per se inherentia sunt necessaria, ergo omnis demonstratio est syllogismus ex per se inherentibus."

10. Here I only say that the series of inherences is the cause of the *knowledge* of the inherence of the predicate in the subject. It will be the case, and this will be seen below, that, in a sense, the series of inherences found in the premises will also be the cause of the *fact* that the predicate inheres in the subject.

11. "Ubicumque enim est necessaria et non accidentalis coherentia predicati cum subiecto, necesse est ut alterum sit egrediens a substantia alterius, aliter enim, destructo altero, nichil prohibet reliquum manere" (130).

12. See pp. 24–31 of this chapter.

13. We will see that this very same language will be used by Grosseteste to understand the relation of universals to particulars. We will turn to this below. We can note, however, that because Grosseteste uses the same language for both per-se predication and universality, the two are in the end one and the same for him.

14. "Habitus itaque eorum in nobis primo est potentialis et materialis passivus et non activus . . ." 403.

15. We will see, e.g., that this is the language Ockham uses in his discussion of *intellectus*. Thomas Aquinas, however, does not use this language in the same way. This could be because of a lack of familiarity with Grosseteste's commentary. The manuscript witnesses listed and discussed in Rossi's introduction to Grosseteste's *Commentarius in Posteriorum Analyticorum Libros,* pp. 31–66, show that it is entirely likely that Grosseteste's commentary did not arrive on the continent until after Aquinas's own commentary was written. It is possible, then, that there were two commentary traditions in the thirteenth century and only one of these traditions was influenced by Grosseteste. However, by the early fourteenth century, scholars at Paris and Oxford alike were referring to both Grosseteste's and Aquinas's commentaries.

16. "Habitus potentialis horum videtur esse sensitiva cognitio. Sensus enim particularis est apprehensivus singularium et sensus communis iudicativus, et est sensus potentia receptiva" (404).

17. The *sensus communis* or "ultimate sense" is discussed by Aristotle in *De Anima,* Bk. III, ch. 2. It is for him a kind of unitary sense that allows an organism to bring together that which has been sensed by various sense organs.

18. Ibid.

19. Ibid. The word "experimentum," which occurs in this passage, cannot be rendered "experiment" without losing the sense of the passage. Grosseteste's point is obviously not that some sort of experiment is being performed and the result is an "experimental universal" (a phrase that is impossible to understand). Rather, he is trying to show how a universal arises out of experience. "*Experimentum*" must mean "experience" but has the sense of the experience of a thing isolated from others. Thus, it is not experience in general, but experience of something.

20. Ibid., 139.

21. Ibid. For a somewhat different gloss on this passage, cf. James McEvoy, *The Philosophy of Robert Grosseteste* (Oxford: Clarendon Press, 1982), pp. 64–67

and pp. 327–29. This latter text, however, agrees with the present interpretation because it shows Grosseteste's mixture of Platonic and Aristotelian ideas about the nature of form.

22. Unfortunately, Grosseteste's extremely interesting light metaphysics cannot be discussed here in great detail. There are several informative accounts of his metaphysics of light, I have found James McEvoy, op. cit., especially helpful. In addition, Hans Blumenberg has undertaken a general study of light as a metaphor for truth in "Licht als Metaphor der Wahrheit. Im Vorfeld der philosophischen Begriffsbildung," *Studium Generale* 10 (1957): 432–47. The role of light in Grosseteste's theory of knowledge is also investigated by Lawrence Lynch, "The Doctrine of Divine Ideas and Illumination in Robert Grosseteste, Bishop of Lincoln," *Medieval Studies* 3 (1941): 163–73. For our purposes, the nature of universals can be explicated without engaging in a debate about the complicated details of this theory and its historical origins. In broad outlines, a metaphysics of light maintains that the universe is created by several contractions and refractions of light, forming various levels of beings as it travels from its source.

23. In fact, even Thomas Aquinas is not averse to talking in terms of "borrowed being," even in his celebrated concept of *esse*. A good example of this is his discussion of the analogy of being in *Summa Contra Gentiles,* Bk. I, chaps. 29–34. For a discussion of this issue in that context, see my "The Analogies of Being in St. Thomas Aquinas," *The Thomist* 58, no. 3 (July 1994): 471–88, esp. 482–87.

24. This argument will be fleshed out in the following chapters.

25. Grosseteste's argument reads: "Item in luce creata, que est intelligentia, est cognitio et descriptio rerum creatarum sequentium ipsam; et intellectus humanus, qui non est ad purum defecatus ita ut possit lucem primam inmediate intueri, multotiens recipit irradiationem a luce creata, que est intellegentia, et in ipsis descriptionibus que sunt <in> intelligentia cognoscit res posteriorem, quarum forme exemplares sunt ille descriptiones" (140).

26. It is the case in almost every such Neoplatonic hierarchical ontology that the top of each ontological level touches the bottom of the level immediately above it. Thus one forms a sort of ontological continuum. Such a continuum is required for the Neoplatonic movement of "procession" from the One and "return" back to the One.

27. Grosseteste, to be sure, does not explicitly ground necessity in universality alone, but in those propositions that are *per se,* universal, and said of their primary subject. These, in the end, turn out to be one and the same condition, as we shall see below.

28. See, for example, p. 111.

Chapter 3

1. "Dicendum sacram doctrinam esse scientiam. Sed sciendum est quod duplex est scientiarum genus. Quaedam enim sunt, quae procedunt ex prin-

cipiis notis lumine naturali intellectus, sicut arithmetica, geometria, et huiusmodi. Quaedam vero sunt, quae procedunt ex principiis notis lumine superioris scientiae, sicut perspectiva procedit ex principiis notificatis per geometriam, et musica ex principiis per arithmeticam notis." *Summa Theologiae,* prima pars, quaestion 1, article 2, hereafter cited as *ST,* followed by part, question, article. Reference to Aquinas's works is more difficult than may seem. While the standard reference should be to Thomas Aquinas, *Sancti Thomae de Aquino Opera Omnia Iussu Leonis XIII P.M. Edita,* edited by Commissio Leonina, 50 vols. (Rome: Commissio Leonina, 1882-), i.e., the so-called Leonine Edition, this edition is not yet finished, and some of the texts that were edited earlier are in need of revision. Furthermore, many publishers have issued Aquinas's texts using the critical text established by the Leonine Edition, but without the critical apparatus. Therefore, I give no separate reference here or in the bibliography for such works unless it seems helpful to the reader. In Thomas Aquinas, *S. Thomae Aquinatis Opera Omnia: Ut Sunt in Indice Thomistico,* edited by Roberto Busa, 7 vols. (Stuttgart-Bad Cannstatt: Frommann-Holzboog, 1980) are included the texts of the Leonine Edition, when available, and for those that have not yet appeared, the editor uses texts that come close to a critical edition, though without critical apparatus. These texts are the basis for his CD-ROM version of the Opera Omnia. Thomas Aquinas, *Thomae Aquinatis Opera Omnia [Computer File]: Cum Hypertextibus in CD-ROM,* edited by Roberto Busa (Milan: Editoria Elettronica Editel, 1992). For the all texts of Aquinas referred to here, I have used the Leonine Edition, but I only give page reference when that would allow easier access.

2. *ST* I, 1, 2 ad 1:"Dicendum quod principia cuiuslibet scientiae vel sunt nota per se, vel reducuntur ad notitiam superioris scientiae."

3. In discussing Aquinas's theory, I use *"per se"* and *"per se nota"* interchangeably at times. This works as long as we have in mind human knowledge. For, as we shall see, there are principles that are *per se*—e.g., "deus est"— that are not *known* to be *per se* by us. In addition, I use "premise" and "principle" interchangeably because a syllogistic premise is a principle, though not all principles are premises in a syllogism. Furthermore, Aristotle also maintains throughout his writings that there may be principles that are central to a syllogism that do not themselves form premises of a syllogism. The principle of noncontradiction would be one such principle.

4. *ST* I, 1, 1c.

5. According to Aristotle, that which is *per se* is indemonstrable. On this topic, see the introduction.

6. *ST* I, 2, 1c.

7. Thomas Aquinas, *Sancti Thomae de Aquino Opera Omnia Iussu Leonis XIII P.M. Edita,* edited by René-Antoine Gauthier O.P., vol. I, pt. 2, *Expositio Libri Posteriorum* (Paris: J.Vrin, 1989), hereafter cited as *Post. Anal.,* followed by book, lectio numbers, and page number. For ease of reference, I also give the paragraph numbers of the Marietti edition. It might be objected

that one ought not to use a literal exposition of Aristotle as a tool to un-pack the *Summa*. This seems to me less problematic for two reasons: (1) Aquinas himself uses terms that are drawn directly from the Aristotelian theory of science; and (2) Aquinas himself uses these senses of *per se* throughout the *Summa*.

8. *Post. Anal.* I, X, 84.

9. *Categories* 2a35: "Everything except primary substances is either said of a subject which is a primary substance or is present in a subject which is a primary substance," Aristotle, *Categories and Propositions,* translated by Hippocrates G. Apostle (Grinnell, Iowa: The Peripatetic Press, 1980), p. 3.

10. There is, in some sense, a problem here. For Aquinas, the only thing that has its own existence itself is God, whose essence is existence. All other beings rely on God for their existence. However, if we place this situation aside for a moment, then substances have their own existence in a way in which accidents do not.

11. *ST* I, 2, 2c: "Unde deum esse, secundum quod non est per se notum quod nos, demonstrabile est per effectus nobis notos."

12. *ST* I, 1, 2c: "Respondeo dicendum, sacram docrtinam esse scientiam. Sed sciendum est, quod duplex est scientiarum genus. Quaedam enim sunt, quae procedunt ex principiis notis lumine naturali intellectus, sicut arithmetica, geometria, et huiusmodi. Quaedam vero sunt, quae procedunt ex principiis notis lumine superioris scientiae: sicut perspectiva procedit ex principiis notificatis per geometriam; et musica ex principiis per arithemeticam notis. Et hoc modo sacra doctrina est scientia, quia procedit ex principiis notis lumine superioris scientiae, quae scilicet est scientia dei, et beatorum."

13. *ST* I, 1, 2c.

14. "*Unde,*" often has a logical function very closely related to "therefore." It makes the claim that what follows the word is a direct consequence of what comes before it.

15. To put this in more Thomistic language, we can say that the question "*an est?*" [does it exist] is prior to the question "*quid est?*" [what is it]. That is to say, while we may not know what kind of thing something is, we certainly know that it exists by the mere fact that we are wondering what it is. If we know what something is, it is thereby implied that we know *that* it is.

16. Throughout this work I use "cognitive state" to translate *habitus* when it is applied to intellectual dispositions.

17. *ST* II-II, 1, 4c.

18. " . . . non enim fides . . . assentit alicui, nisi quia est a deo revelata. . . ."

19. We will turn to a detailed discussion of Aquinas's position on *intellectus* below.

20. *ST* I, 14, 1c: "Unde et supra diximus, quod formae, secundum qoud sunt magis immateriales, secundum hoc magis accedunt ad quandam infinitatem. Patet igitur, quod immaterialitas alicujus rei est ratio, quod sit cognoscitiva, et secundum modum immaterialitatis est modus cognitionis."

21. *ST* I, 14, 1c.

22. *ST* I, 14, 1c: "Unde, cum deus sit in summo immaterialitatis . . . sequitur, quod ipse sit in summo cognitionis."

23. *ST,* I, 14, 8 ad.3. I have translated "*mediae*" in a deliberately ambiguous way, for it is also ambiguous in the Latin. Does Aquinas mean that it is middle as a middle term in a syllogism, i.e., that which allows one to move from knowledge of one to knowledge of the other? Or does Aquinas mean that the things occupy the midway point between our knowledge and the knowledge that God has? My interpretation of this term will become obvious in a moment.

24. See note 11 above for Aquinas's text.

25. In medieval terminology, the most proper kind of demonstration is *propter quid*. This kind of demonstration gives the reason for a fact. There are, however, also demonstrations *quia,* which do not supply the reason for the fact, but merely demonstrate the fact itself. Aquinas argues, then, that the proofs for the existence of God are not *propter quid,* but rather *quia.* But this lesser form of scientific syllogism is still a *scientific* syllogism.

26. As a preface to his proofs for the existence of God, Aquinas states that effects can be used "in loco definitionis" (*ST* I, 2, 2c). This does not apply merely to God, but it applies to the greatest extent to God.

27. *ST* I, 19, 1c: "Voluntas enim intellectum consequitur."

28. *ST* I, 19, 8c: "Sic enim scientia dei se habet ad omnes res creatas, sicut scientia artificis se habet ad artificiata."

29. Ibid.: "Unde oportet, quod forma intellectus sit principium operationis: sicut calor est principium calefactionis; sed considerandum est, quod forma naturalis, inquantum est forma manens in eo, cui dat esse, non nominat principium actionis, sed secundum quod habet inclinationem ad effectum; et similiter forma intelligibilis non nominat principium actionis, secundum quod est tantum in intelligente; nisi adjungatur ei inclinatio ad effectum, quae est per voluntatem."

30. *ST* I, 19, 3c: "Unde, cum bonitas dei sit perfecta, et esse possit sine aliis, cum nihil ei perfectionis ex aliis accrescat, sequitur, quod alia a se eum velle non sit necessarium absolute, et tamen necessarium est ex suppositione. Supposito, enim, quod velit, non potest non velle, quia non potest voluntas ejus mutari."

31. The relation between the unique first cause and the unique world works in both directions. At *ST* I, 11, 3c, Aquinas argues that from the order of all things, we can discover that there is one, unique, first cause of that order. At *ST* I, 68, 3c, he argues that from the fact that there is one, unique, first cause we can discover that there is only one world and it is one on account of its order.

32. *ST* I, 68, 3c: "Et ideo illi potuerunt ponere plures mundos, qui causam mundi non posuerunt aliquam sapientiam ordinatem, sed casum. . . ."

33. Throughout this text terms such as "natural state," "in this state," and "in our current state," are all translations of the Latin terms "pro statu isto," or

"viatoris." These terms refer to human beings in the state in which they are after the fall, but not yet having died and, receiving grace, enjoying the beatific vision (i.e., seeing "the face of God"). The blessed [*beatus*] are distinguished from us in our current state only by the fact that the blessed see the essence of God, while we do not and cannot—without violating our nature in this state.

34. While Aquinas makes no distinction between "our theology" and "theology itself," such a distinction is inherent in his idea of theology and the need for the knowledge that the blessed have. Furthermore Aquinas raises the issue at *ST* I, 1, 5, ad 1 that something could be more certain "according to nature" and yet less certain for us on account of a debility of our intellect. This would open the possibility that theology in itself ("according to nature") would have a different character than our theology. This separation between "our theology" and "theology itself" will be made explicit by Duns Scotus (see chapter 4 below). It is clear, however, that such a distinction is made necessary here by Aquinas.

35. Aquinas often distinguishes between these two kinds of vision (*ST* I, 67, 1c). At other times, however, he adds a third member these types of vision: imagination (*ST* I, 93, 6 ad.4; II-II, 174, 1 ad. 3; II-II; 175, 3 ad.4).

36. *ST* I, 67, 1c. The whole argument reads: " . . . sicut patet in nomine visionis, quod primo impositum est ad significandum actum sensus visus; sed propter dignitatem, et certitudinem hujus sensus extensum est hoc nomen, secundum usum loquentium, ad omnem cognitionem aliorum sensuum."

Aquinas goes on to discuss this same kind of extension with regard to the concept of light: "Nam primo quidem est institutum ad significandum id, quod facit manifestationem in sensu visus; postmodum autem extensum ad significandum omne illud, quod facit manifestationem secundum quamcumque cognitionem."

37. *ST* I, 12, 1c: "In ipso enim est ultima perfectio rationalis creaturae, quod est ei principium essendi." Notice here that Aquinas argues that the highest perfection of a rational creature *does not* consist in grasping being, but rather in the principle or ground of being. Being itself must be brought under the domain of its own rational ground.

38. *ST,* II-II, 1, 5c.

39. This whole process will be analyzed by Aquinas in his commentary on *Posterior Analytics.* We will turn to this text below.

40. *Post. Anal., prooemium:* "Tercius uero actus rationis est secundum id quod est proprium rationis, scilicet discurrere ab uno in aliud, ut per id quod est notum deueniat in cognitionem ignoti. . . ."

41. *ST* I, 19, 5c.

42. On the question of the principles of theology, cf. Ulrich Köpf, *Die Anfänge der Theologischen Wissenschaftstheorie Im 13. Jahrhundert* (Tübingen: J. C. B. Mohr, 1974), pp. 142ff. Köpf there traces the origin of the comparison of the articles of faith to the principles of science to either Odo Rigaldi or an anonymous author of the text appearing in Vat. Lat. 782. This means that

the problem that Aquinas tackles here precedes him by quite some time. The solution to this problem of the principles that Aquinas offers (subalternation), however, does not have such an old history. Cf. pp. 145–49.

43. Herveus Natalis raises an argument very similar to this one in his *Commentary* on the *Sentences:* " . . . non plus repugnat obscuritas fidei scientiae de credibilibus, quam obscuritas phantasmatis scientiae de sensibilibus. Sed non obstante obscuritate phantasmatis habetur certa scientia de sensibilibus. Ergo non obstante obscuritate fidei potest haberi scientia de credibilibus." Herveus Natalis, *In Quatuor Libros Sententiarum Commentaria* (Paris: Dyonisius Moreau & Son, 1647), fo. 3. Herveus ultimately refutes this view that comes in a section that discusses Henry of Ghent's *lumen medium*.

44. It should be noted that post-Reformation Catholic thinkers could not reject Aquinas's solution. Their only recourse, then, was to attempt to make sense of it in the face of more than two centuries of arguments against it. Bañez, too, struggles with this solution. His only hope is to show that whatever its faults, theology must be a science. See Dominico Bañes, *Scholastica Commentaria in Primam Partem Summae Theologicae s. Thomae Aquinatis* (Madrid: Editorial F.C.D.A., 1934), 1, 1.2.

45. Köpf, *Die Anfänge der Theologischen Wissenschaftstheorie im 13. Jahrhundert*, p. 139–40.

46. Thomas Aquinas, *Sancti Thomae de Aquino Opera Omnia Iussu Leonis XIII P.M. Edita*, edited by René-Antoine Gauthier O.P., vol. I, pt. 1, *Expositio Libri Peryermenias* (Paris: J. Vrin, 1989), *prooemium.;* hereafter cited as *Per. Herm.*, followed by lectio number.

47. The composition of simples would lead to an affirmative proposition, e.g., 'Socrates is white', while the division would lead to a negative proposition, e.g., 'Socrates is not white'. Aquinas states this more clearly : " . . . de his uero, que pertinent ad secundam operationem, scilicet de enunciatione affirmatiua et negatiua. . . ."

48. Ibid.: "Additur autem et tercia operatio, scilicet ratiocinandi, secundum quod ratio procedit a notis ad inquisitionem ignotorum."

49. Ibid.

50. In fact, as we will see below, this third act is syllogistic. This means that two known propositions are required. There are cases, it seems, in which I can move from the knowledge of just one proposition to the knowledge of another. Here, the theory of "consequences" does not come into play because "consequences" were most often seen to be syllogisms in a shortened form. Rather, the move from effect to cause seems to be warranted in certain cases, e.g., "This is smoke, therefore there is fire."

51. Ibid. The entire argument reads: "Secunda uero ordinatur ad terciam: quia uidelicet oportet quod ex aliquo uero cognito, cui intellectus assenciat, procedatur ad certitudinem accipiendam de aliquibus ignotis."

52. Here Aquinas expands the formulation of the *prooemium* to *Per. Herm*. In that text, he called it "*intelligencia indiuisibilium*." Here, however, he calls it "*intelligencia indiuisibilium siue incomplexorum*," *Post. Anal., prooem*.

53. *Post. Anal., prooemium:* "Pars autem logicae que primo deseruit processui pars iudicatiua dicitur, eo quod iudicium est cum certitudine scientiae. . . ."

54. Ibid.

55. This explication is found in *Post. Anal.,* II, 20.

56. Based on the divisions of the *prooemia* to both *Per. Herm.* and *Post. Anal.* we can give "reason" a precise definition. Reason is the name given to the operation of the intellect that goes from the knowledge of one thing to the knowledge of another thing.

57. Ibid., p. 245: "Ratio autem non sisit in experimento particularium, set ex multis particularibus in quibus expertus est, accipit unum commune. . . ."

58. Ibid.

59. *ST* II-II, 1, 4c: "Respondeo dicendum quod fides importat assensum intellectus ad id quod creditur. Assentit autem alicui intellectus dupliciter. Uno modo, quia ab hoc movetur ab ipso obiecto, quod est vel per seipsum cognitum, sicut patet in principiis primis, quorum est intellectus; vel est per aliud cognitum, sicut patet de conclusionibus, quarum est scientia. Alio modo intellectus assentit alicui non quia sufficienter moveatur ab obiecto proprio, sed per quandam electionem voluntarie declinans in unam partem magis quam in aliam. Et si quidem hoc fit cum dubitatione et formidine alterius partis, erit opinio; si autem fit cum certitudine absque tali formidine, erit fides. Illa autem videri dicitur quae per seipsa movent intellectum nostrum vel sensum ad sui cognitionem."

60. Ibid., II-II, 1, 5c.

61. Such strange combinations of diverse lexical histories happens frequently in medieval thought. More often than not, it comes from the hermeneutic principle that all authorities must be reconciled. This means that such coincidences in the use of terms often become the springboard for reconciliation. Diverse semantic histories never prevented reconciliation through this method. This is one area that would provide a fruitful understanding of the relation between theology and philosophy in the Middle Ages. Often, philosophical concepts come to have theological significance because of the historical accident of translation of the biblical text into Latin.

62. While there can be *intellectus* of that which is accidental, as Aquinas has shown, such *intellectus* does not pertain to science. It is only the *intellectus* of universal natures, i.e., essences, that pertains to science.

63. While Aquinas and all other medieval commentators on *Posterior Analytics* use the term "sense," which would cover all the senses, the examples they use rely only on vision and not on the other senses.

64. It is certainly beyond the scope of my argument to engage the tremendous body of literature that has arisen on this subject. My intent here is rather to show how, *in spite of his own position,* Aquinas can be seen as stepping beyond the limits of both negative theology and analogy.

65. *ST* I, 3, *prooemium.*

66. Ibid., " . . . quia nullum corpus movet non motum . . . Ostensum est autem supra (*ST* I, 2, 3c) quod Deus est primum movens immobile: unde manifestum est, quod Deus non est corpus."

67. Ibid., " . . . quia necesse est id, quod primum ens, esse in actu, et nullo modo in potentia. Licet enim in uno, et eodem, quod exit de potentia in actum, prius sit potentia, quam actus, tempore; simpliciter tamen actus prior est potentia, quia quod est in potentia, non reducitur in actum, nisi per ens actu. Ostensum est autem supra (*ST* I, 2, 3c) quod deus est primum ens. Impossible est igitur, quod in Deo sit aliquid in potentia; omne autem corpus est in potentia, quia continuum, inquantum hujusmodi, divisible est in infinitum: impossibile est igitur Deum esse corpus."

68. Ibid., " . . . quia Deus est id, quod est nobilissimum in entibus, ut ex dicitis patet (*ST* I, 2, 3c) impossibile est autem aliquod corpus esse nobilissimum in entibus, quia corpus aut est vivum, aut non vivum. Corpus autem vivum, manifestum est, quod est nobilius corpore non vivo: corpus autem vivum non vivit, inquantum corpus, quia sic comne corpus viveret: oportet igitur, quod vivat per aliquid aliud, sicut corpus nostrum vivit per animam. Illud autem, per quod vivit corpus, est nobilius, quam corpus: impossibile est igitur Deum esse corpus."

69. In this sense, the proofs for the existence of God do not merely prove *that* God is, but also prove something significant about God (i.e., that God is prime mover, most noble, actual, etc.). These positive features are then used to deny corporeal composition of God in the next question.

70. *ST* I, 13, 5c.

71. Aristotle goes so far as to assert that of particulars there is no knowledge, only sense. This makes the relation between sensation and knowledge all the more problematic. We will turn to this assertion in chapter 5.

72. This seems to be the position of the Condemnations of 1277, at any rate.

Chapter 4

1. Propositions stemming from natural philosophy and cosmology form by far the largest number of propositions condemned. Many of these propositions directly refer to what God or the "first cause" can and cannot do, see, e.g., 34, 35, 37, 38, 42–50, *Chartularium Universitatis Parisiensis,* edited by Henricus Denifle, O.P. (Paris: Culture et Civilisation, 1964), pp. 545–46. Alain de Libera has placed these condemnations within the context of the general struggle of medieval thought in relation to the university in his *Penser Au Moyen Âge* (Paris: Éditions du Seuil, 1991), esp. pp. 143–48. While I am in general inclined to agree with his conclusions there, I am here more interested in the *effects* that the condemnations had on subsequent thinkers, who most certainly could not call into question the legitimacy of them by pointing to the circumstances of their origin and construction.

2. Proposition 34, "Quod prima causa non posset plures mundos facere." *Chartularium Universitatis Parisiensis,* p. 545.

3. My argument here is *not* that the commission of theologians was concerned about the loss of the existing singular in the face of the search for rational ground. Rather, I am simply pointing out that the condemnations had the effect that the rational ground could not bind God's creative power. The result of this is that the singular arises without rational ground in many philosophers after the condemnations. That being said, one could point out the concern of the commission to safeguard also the singularity of God such that God does not have a rational ground. If, then, God is the cause of all things, existing singulars will also be without rational ground. Thus, it could be shown that the commission was concerned precisely with this question.

4. Thomas Aquinas, *Summa Theologiae,* prima pars, question 97, article 3.

5. Ibid.: "Et ideo illi potuerunt ponere plures mundos, qui causam mundi non posuerunt aliquam sapientiam ordinantem, sed casum."

6. For those, like Etienne Gilson, who claim that the "Thomistic synthesis" achieved a reconciliation of God's creative power with God's knowing intellect, the condemnations must be read as a misinterpretation of Thomistic thought, or were the result of radical and irreligious "Averroists," or as a result of rivalry between mendicants and the secular faculty at Paris (and eventually at Oxford) or as a result of rivalry between the mendicant orders themselves. My interest here is not in solving that question but in showing how Aquinas could be read in this way and what the implications are for subsequent thinkers. However, I think that Aquinas could achieve his "synthesis" only through subsuming one side to the other. I would argue that the notion of "order" in Aquinas functions so as to bring God's creative will under the divine intellect.

7. Scotus often treats knowledge of existing singulars *as existing* under the heading of "intuitive knowledge." However, his theory of science as expounded in his various commentaries on *Sentences* does not seem to *require* intuitive knowledge at all. This topic will be taken up below.

8. For a tracing of this shift, see Edward O'Connor, "The Scientific Character of Theology According to Scotus," in *De Doctrina Ioannis Duns Scoti,* edited by Commisio Scotistica (Rome: Commisio Scotistica, 1968), p. 9.

9. Ioannis Duns Scotus, *Opera Omnia,* edited by Commissio Scotisticae, vol. 1, (Vatican City: Typis Polyglottis Vaticanis, 1950-), p. 100: "proportio obiecti ad potentiam est proportio motivi ad mobile vel activi ad passivum; proportio obiecti ad habitum est sicut proportio causae ad effectum." Where available, I use the Vatican City edition of Scotus's works. The *Ordinatio* will be cited as *Ord.,* followed by paragraph number, with the volume and page number in parentheses, e.g., *Ord.,* n. 24 (Vat. 1: 100).

10. Ludger Honnefelder, *Ens in Quantum Ens: Der Begriff Des Seienden Als Solchen Als Gegenstand der Metaphysik Nach der Lehre Des Johannes Duns Scot* (Münster: Aschedendorff, 1979), p. 7.

11. 71b10: "We think we know each [thing] without qualification, but not in the sophistical manner with respect to an attribute, when we think that (a) we know the cause through which the thing exists as being the cause of that thing and that (b) the thing cannot be other than what it is." Aristotle, *Posterior Analytics*, translated by Hippocrates Apostle (Grinnell, Iowa: Peripatetic Press, 1981), p. 2. See. *Ord.*, n. 211 (Vat. 1: 144), where Scotus argues that the perfection of science lies in certitude, not in explanation. Here we can see a trend that moves away from Aquinas and Grosseteste. Aquinas could not make this move completely because of the crucial role that forms as causes play in the order of the universe. For Aquinas, then, necessity, explanation, and certainty are all inextricably linked. This move away from necessity and toward certainty is charted by O'Connor, op. cit., p. 17.

12. Ioannis Duns Scotus, *Opera Omnia*, edited by L. Vives, 26 vols. (Paris, 1891–1895), n. 24. This work will be cited as *Rep.*, followed by paragraph number with volume and page in parentheses, e.g., *Rep.*, n. 24 (Vives 1: 203). See also *Ord.*, prol., pars IV, q. 1&2, n. 208 (Vat. 1): " . . . dico quod scientia stricte sumpta quattuor includit, videlicet: quod sit cognitio certa, absque deceptione et dubitatione; secundo, quod sit de cognito necessario; tertio, quod sit causata a causa evidente intellectui; quarto, quod sit applicata ad cognitum per syllogismum vel discursum syllogisticum."

13. Here, like we saw in Aquinas, Aristotle's *De Anima* is the source for these operations of the intellect.

14. Ioannis Duns Scotus, *Quaestiones super Libros Metaphysicorum Aristotelis. Libri I-V,* edited by R. Andrews, and others, Opera Philosophica, 3 (St. Bonaventure: The Franciscan Institute, 1997), p. 99–100: "Igitur nullo actu intellectus cognoscitur aliquid a nobis nisi praecesserit cognitio sensibilium in sensu." This text will be cited as *Quaest. in Metaph.*, followed by volume and page number.

15. Ibid.

16. On this topic, see Honnefelder, op. cit., pp. 170ff.

17. I use here the ambiguous phrase "pertains to" in order to leave aside the question of just how it is that truth is related to a proposition. This question will take on increasing importance in medieval philosophy with the two main positions taking the dominant role. On the one hand, there were philosophers who, following Aegidius Romanus, argued that a proposition signifies a "complex significable," and the truth of the proposition belongs to the relation between that proposition and this complex significable. On the other hand, there were philosophers who argued that truth is a property of a true proposition and that the complex significable was a superfluous notion.

18. *Quaest. in Metaph.*, 3: 108: "Quia ad quamcumque apprehensionem sensitivam imprimuntur intellectui ens et res. Simplicibus apprehensis a sensu vero vel falso, propositiones fiunt virtute propria intellectus: primo de universalibus, postea de aliis. De universalissimis, factis communibus conceptionibus, statim intellectus illis assentit, non propter sensum, immo certius

quam posset per sensum, dato quod a sensu accepisset cognitionem veritatis illarum propositionum. Alias propositiones facit et immediatas de minus universalibus, sed non statim notas nec scitas esse immediatas, quia termini non cognoscuntur."

19. In this text, Scotus often uses "sensitiva" and "apprehensiva" interchangeably. We will see that Ockham is going to distinguish between the two such that there can be a "sensitive" act of apprehension *and* an "intellective" act of apprehension. This forces him to push the act of judging back one step in the process.

20. *Quaest. in Metaph.*, 3: 110: " . . . tum quia de omni sensitiva potest intellectiva iudicare qualis ipsa est. Tum quia certitudo numquam est in apprehendendo verum nisi talis sciat veritatem apprehensam, vel sciat illud esse verum quod apprehendit. . . . Tum quia numquam sensus percipit immutabilitatem obiecti, licet illud, quod immutabile est, sentiat. . . ."

21. *Quaest. in Metaph.*, 3: 224: "Notandum quod in sensu est una cognitio intuitiva, primo propria, alia primo et per se propria per speciem, sed non intuitiva."

22. Ibid., 3: 230.

23. Scotus provides many definitions of this two forms of cognition or knowledge. See, e.g., *Quodlibet, Q.* 14, n. 10 (Vives 26: 39), *Ord.* II, d. 3, pars 2, q. 2 nn. 318ff (Vat. 7: 552f).

24. Honnefelder, op. cit., p. 219.

25. *Quaest. in Metaph.*, 3: 231: "Quia quidquid perfectionis simpliciter est in inferiori, videtur ponendum in superiori. Perfectionis simpliciter est in sensitiva cognitione quod cognoscit aliquid in quantum praesens est per essentiam; ergo huiusmodi cognitio videtur hic competere intellectui."

26. It has been shown by many commentators that for Scotus *visio* and *cognitio intuitiva* are often synonymous. See Allan B. Wolter, "Duns Scotus on Intuition, Memory and Our Knowledge of Individuals," in *History in the Making,* edited by Linus J. Thro (New York: University Press of America, 1982), pp. 81–104; and Stephen D. Dumont, "Theology as a Science and Duns Scotus's Distinction Between Intuitive and Abstractive Cognition," *Speculum* 64 (1989): 579–99.

27. See, for example, *Ord.* II, n. 321 (Vat. 7); *Ord.* I, n. 35–37 (Vat. 1).

28. Wolter, "Scotus on Intuition," pp. 81–2, states "At the intellectual level, [intuitive knowledge] is an act of simple awareness or intelligence in which some object is grasped holistically [*simul totum*] as present and existing here and now. Hence it is not to be confused with the subsequent contingent or existential judgments that explicate its conceptual content."

29. There has been some controversy, both in the Middle Ages and today, about whether Scotus admits the possibility of intellective (as opposed to sensitive) intuitive knowledge of singulars. Indeed, in his commentary on *Metaphysics,* he seems to hold both positions at once (see, e.g., II, q. 3, p. 224 as opposed to p. 232; see also VII, q. 15). This issue does not need to be resolved here, but I am convinced by Honnefelder, op. cit., p. 232ff.,

who brings texts from both the *Ordinatio* and the *Metaphysics* commentary, as well as Wolter, "Scotus on Intuition," pp. 81–104, who uses this precisely as a way to chart the stages of Scotus's development on this topic. What is clear, however, is that Scotus maintains, on the one hand, that such intuitive knowledge is not required as the basis of *scientia,* and on the other hand, though less so, that intellective intuitive knowledge is not the basis of *scientia.* I will return to this latter point below.

30. *Rep.* (Vives 22: 41).

31. Dumont, op. cit., p. 584.

32. We will see in the following chapter that this is one point with which Ockham will take issue. For him, all abstractive knowledge must have as a partial cause intuitive knowledge of the same thing.

33. *Ord.* III, d. 3 (Vat. 7: 403): "Non solum autem ipsa natura de se est indifferens ad esse in intellectu et in particulari, ac per hoc et ad esse universale et particulare (sive singulare). . . ."

34. Ibid., p. 416: " . . . necesse est per aliquid positivum intrinsecum . . . et illud positivum erit illud quod dicetur esse per se causa individuationis. . . ."

35. Ioannis Duns Scotus, *Quaestiones super Libros Metaphysicorum Aristotelis. Libri VI-IX,* edited by R. Andrews, and others, Opera Philosophica, 4 (St. Bonaventure: The Franciscan Institute, 1997), pp. 300–301: " . . . intellectus noster in hoc statu non intelligit per se singulare, nec sensus sentit."

36. Ibid., 4: 302: "Nulla potentia cognoscitiva in nobis cognoscit rem secundum absolutam suam cognoscibilitatem, in quantum scilicet est in se manifesta, sed solum in quantum est motiva potentiae. Quia cognotivae hic moventur ab obiectis; natura autem non movet secundum gradum singularitatis."

37. Ibid., 4: 303: "Ex hoc sequitur quod intellectus, immediate receptivus actionis obiecti, potest moveri a singularitate; non autem qui est receptivus mediante actione naturali. Tantum primus est intellectus angelicus, qui videt immediate singulare materiale. Secundus est noster intellectus, in quem non agit natura nisi mediante in gignitione in sensum, quae potest dici actio naturalis materialis, respectu illius quae est intelligibilis, in intellectum."

38. *Ord.* IV, d. 45, q. 2, n. 12 (Vives 21: 305): "Talis autem cognitio, quae dicitur intuitiva, potest esse intellectiva, alioquin intellectus non esset certus de aliqua existentia alicuius obiecti."

39. *Ord.,* prol., pars III, q. 3, n. 141 (Vat. 1).

40. Ibid., n. 142.

41. Ibid., n. 144.

42. Ibid. n. 145: "Ille habitus qui dicitur scientia est species intelligibilis primi obiecti. . . ." See also O'Connor, op. cit., 6.

43. E.g., *Rep.* (Vives 22: 9a); Ioannis Duns Scotus, *Treatise on God as First Principle,* translated by Allan B. Wolter O.F.M. (Chicago: Franciscan Herald Press, 1966), p. 5. In this latter text, the kind of essential order we are discussing seems to be that which is called the "order of eminence." I will return to this point presently.

44. *Rep.* (Vives 22: 9a).
45. Ibid., (Vives 25: 35a). See also Stephen D. Dumont, "The *Propositio Famosa Scoti:* Duns Scotus and Ockham on the Possibility of a Science of Theology," *Dialogue* 31 (1992): 419.
46. Dumont, "*Propositio Famosa,*" 419.
47. In this, I agree with the analysis of Dumont.
48. In this condition, Scotus seems to accept Henry of Ghent's position that something is added to our knowledge through discourse alone, without that thing having to be "intuited" or "seen." In fact, as Dumont (op. cit.) argues, Scotus's conception of abstractive knowledge seems to fill precisely the same role as Henry's *intellectus.* My point is that when this is the case, the singularity of the singular (or the singular *qua singular*) remains untouched by science.
49. *Ord.* Prol. Pars III, q. 3, n. 145 (Vat. 1).

Chapter 5

1. One often finds attempts to do just that. A good example comes from Marilyn McCord Adams, *William Ockham* (Notre Dame: University of Notre Dame Press, 1987), p. 3: "Ockham's philosophical focus, whether he is doing logic, natural science, or theology, is on the branch of metaphysics commonly called 'ontology'."
2. On these facets of Ockham's thought and their consequences, cf. Heiko A. Oberman, "*Via Antiqua* and *Via Moderna:* Late Medieval Prolegomena to Early Reformation Thought," in *From Ockham to Wyclif,* edited by Anne Hudson and Michael Wilks (Oxford: Basil Blackwell, 1987), pp. 445–63. Oberman approaches Ockham and late medieval thought from the point of view of the Reformation. While this may possibly open his understanding of Ockham to some faults, Oberman is consistently able to avoid these. The reason for this, to my mind, is that he avoids any questions about the state of mind of Ockham (and other late medieval thinkers) and shows, instead, what the consequences of their thought is, *whatever they intended to the contrary.* What matters for us is not what Ockham thought he was doing (which we will never know and which is the least interesting of all questions) but what he in fact said.
3. William of Ockham, *Scriptum in Librum Primum Sententiarum Ordinatio,* edited by Gedeon Gál, O.F.M., and Stephen Brown, O.F.M., Guillelmi de Ockham Opera Philosophica et Theologica, OT I (St. Bonaventure: Franciscan Institute Publications, 1967), p. 87–88. Hereafter cited as *I Sent.,* followed by page number. The formulation comes directly from Scotus.
4. *I Sent.,* 88: "Per primam condicionem excluditur et opinio et suspicio et fides et huiusmodi, quia nulla illarum est evidens. Per secundam excluditur evidens notitia contingentium quae non est scientia proprie dicta, quia non est veri necessarii. Per tertiam condicionem excluditur notitia evidens primorum principiorum, quia illa non potest haberi per discursum syllogisticum."

5. A proposition is able to be caused by premises if and only if there is a middle term through which the proposition can be demonstrated in a syllogism that follows the rules laid down in *Posterior Analytics*.

6. *I Sent.*, 88: "Dico autem 'nata causari', quia non est necessarium quod de facto causetur per tales premissas, quia potest per experientiam causari."

7. This would apply to both the question of one and the same truth being known by two sciences and the question of whether one and the same truth can be known by two completely different habits. The first question seems to cause little problem, as it was generally agreed that theology includes many truths that are also part of metaphysics. The second does not cause problems unless one wants to argue that a truth known by experience is not known to any lesser extent than a truth known by science. Such a position follows immediately when one argues that *scientia* is not about explanation but about certainty. If that is the case, then I can know any number of truths with certainty that are not demonstrated. While he is not so explicit here, Ockham's position does tend to weaken the strong position that *scientia* held over experience.

8. Though even when it is made evident by a demonstration, that demonstration will rely on premises that will have to be made evident outside of any demonstration—i.e., in experience.

9. *I Sent.*, 83: " . . . dico quod non omnis passio est demonstrabilis de suo subiecto, sed aliqua passio non potest sciri inesse suo subiecto nisi praecise per experientiam, et nullo modo per demonstrationem. Et hoc non tantum est verum de propositione per se nota in qua praedicatur passio de sua subiecto, sed etiam de propositione frequenter non per se nota, sicut de ista 'calor est calefactivus', et de similibus."

10. *I Sent.*, 76.

11. *I Sent.*, 76–77: " . . . quod sit 'propositio dubitabilis', patet: quia per hoc excluditur propositio per se nota quae quamvis sit necessaria et possit esse evidenter nota, quia tamen non est dubitabilis ideo non est scibilis scientia proprie dicta. . . . Igitur omnis propositio scibilis est primo dubia vel apparet falsa, et postea per principia manifestatur veritas eius."

12. Maier has interpreted "*propter quid*" as "metaphysical explanation," and "*quia*" as "phenomenological description." Anneliese Maier, *Die Vorläufer Galileis* (Rome: Edizioni di Storia e Letteratura, 1949); translated in Anneliese Maier, *On the Threshold of Exact Science: Selected Writings of Anneliese Maier on Late Medieval Natural Philosophy,* edited and translated by Steven Sargent (Philadelphia: University of Pennsylvania Press, 1982). This interpretation is interesting for several reasons. The most important of these is that the later medieval period saw a rise in the interest in theory of *demonstratio quia* and an attempt to put this on more secure scientific footing. This would mean, in her terms, that there is a move away from metaphysical explanation toward phenomenological description. If this were true, it would support the thesis for which I am arguing here. If we borrow this language, we can say that Ockham moves science in general, and not just *quia,* from

a search for metaphysical explanation toward phenomenological description, that is, away from providing a rational ground and toward its relation to experience.

13. *I Sent.*, 109–10.
14. *I Sent.*, 110.
15. *I Sent.*, 111: " . . . nihil intrinsecum deo potest de divina essentia demonstrari ita quod divina essentia in se subiciatur et aliquid quod est realiter divina essentia praedicetur in se."
16. On the subject of demonstration in general see Damascene Webering, O.F.M., *Theory of Demonstration According to William of Ockham* (St. Bonaventure: Franciscan Institute Publications, 1953). On demonstrating theological propositions in particular, see pp. 135–41.
17. *I Sent.*, 113: "Dico quod nulli habenti illos terminos potest esse dubia, sive cognoscantur abstractive sive intuitive, quia sive sic sive sic cognoscatur divina essentia, necessario distincte cognoscitur essentia divina."
18. On the notion of "immediate" in Aristotle's *Posterior Analytics*, see ch. 1.
19. *I Sent.*, 114.
20. *I Sent.*, 115: "Quarta conclusio est: quod conceptus connotativi et negativi communes deo et creaturis possunt de divina essentia demonstrari. . . ."
21. See Webering, op. cit., 139.
22. See William of Ockham, *Summa Logicae,* edited by Philotheus Boehner, O.F.M., Gedeon Gál, O.F.M. and Stephan Brown, Guillemi de Ockham Opera Philosophica et Theologica, OP I (St. Bonaventure: Franciscan Institute Publications, 1974), p. 36; hereafter cited as *SL,* followed by page number.
23. *SL,* 36: "Et tale nomen proprie habet definitionem exprimentem quid nominis. . . . Sicut est de hoc nomine 'album', nam 'album' habet definitionem exprimentem quid nominis . . . 'aliquid informatum albedine' vel 'aliquid habens albedinem'."
24. *I Sent.*, 116: " . . . quod conceptus connotativi et negativi proprii Deo non sunt de divina essentia in se demonstrabiles a priori." It should be noted that there are several classes of demonstration that are arranged in a hierarchical order from lowest to highest. Therefore, while some propositions may be demonstrated a posteriori, such a demonstration is not the highest form of demonstration. Similarly, a demonstration that is a priori, but not universal, is not the highest form of demonstration. The highest form of demonstration is usually referred to as "*demonstratio potissima,*" and refers to an a priori, *propter quid,* universal demonstration.
25. While there are some difficulties in particular places, the account of Webering, op. cit., is still largely useful. For a somewhat different interpretation, also not without some problems in particular areas, see Ernest Moody, *The Logic of William of Ockham* (New York: Sheed & Ward, 1935).
26. The best discussions of the subject can be found in Jürgen Miethke, *Ockhams Weg zur Social Philosophie* (Berlin: Walter de Gruyter, 1969); Helmar Junghans, *Ockham im Lichte der neueren Forschung* (Hamburg: Lutherisches

Verlagshaus, 1968), esp. pp. 163ff; Marilyn McCord Adams, "Intuitive Cognition, Certainty and Skepticism in William Ockham," *Traditio* 26 (1970): 389–98; André Goddu, "The Dialectic of Certitude and Demonstrability According to William of Ockham and the Conceptual Relation of His Account to Later Developments," in *Studies in Medieval Natural Philosophy,* edited by Stefano Caroti (Firenze: L. S. Olschki, 1989), pp. 95–131; Paul A. Streveler, "Ockham and His Critics on: Intuitive Cognition," *Franciscan Studies* 35 (1975): 223–36 (though I think he misconstrues the point of Adams in the article cited above); T. K. Scott, "Ockham on Evidence, Necessity and Intuition," *Journal of the History of Philosophy* 9 (1971): 15–41; John F. Boler, "Ockham on Intuitive Cognition," *Journal of the History of Philosophy* 11 (1973): 95–106; and item, "Ockham on Evident Cognition," *Franciscan Studies* 36 (1976): 85–98; Marilyn McCord Adams, *William Ockham* (Notre Dame: University of Notre Dame Press, 1987); Katherine Tachau, *Vision and Certitude in the Age of Ockham: Optics, Epistemology and the Foundations of Semantics 1250–1345,* Studien und Texte Zur Geistesgeschichte Des Mittelalters, vol. XXII (Leiden: E. J. Brill, 1988); Rega Wood, "Intuitive Cognition and Divine Omnipotence: Ockham in Fourteenth-Century Perspective," in *From Ockham to Wyclif,* edited by Anne Hudson and Michael Wilks (London: Basil Blackwell, 1987), pp. 51–61 (this text provides a catalogue of various fourteenth century thinkers' theories of intuitive cognition). Finally, Gordon Leff, *William of Ockham: The Metamorphosis of Scholastic Discourse* (Manchester: Manchester University Press, 1975), is almost completely unreliable. The text is filled with errors—including errors in translations—contradictions, and misinterpretations. This text should be used only with extreme caution.

27. This question has been put to rest in its initial form as raised by Etienne Gilson, "The Road to Skepticism," in *The Unity of Philosophical Experience* (New York: Charles Scribner's and Sons, 1937) and Anton Pegis, "Concerning William of Ockham," *Franciscan Studies* 2 (1944): 465–80. However, it has resurfaced in several forms that seem more trenchant.

28. *I Sent.,* 76.

29. In the *SL,* Ockham follows Grosseteste in positing only two modes of per-se predication: (1) when the predicate defines the subject (i.e., when the predicate is included in the definition of the subject) or something superior to the subject (e.g., "every human is composed of matter and form"); (2) when the subject, or something superior to the subject, defines the predicate or something inferior to the predicate. Ockham removes all talk of causality—or at least interprets causality as a relation among terms and not of things. For Ockham, these modes of per-se predication mean, in the end, that the one term "imports or signifies something which is signified by the other," *SL,* 517f.

30. *I Sent.,* 83; *SL,* 522–24.

31. *I Sent.,* 85.

32. *I Sent.*, 16: " . . . inter actus intellectus sunt duo actus quorum unus est apprehensivus, et est respectu cuiuslibet quod potest terminare actum potentiae intellectivae, sive sit complexum sive incomplexum. . . ."

33. *I Sent.*, 16: "Alius actus potest dici iudicativus quo intellectus non tantum apprehendit obiectum sed etiam illi assentit vel dissentit. Et iste actus est tantum respectu complexi, quia nulli assentimus per intellectum nisi quod verum reputamus, nec dissentimus nisi quod falsum aestimamus." The fact that truth and falsity belong only to propositions follows directly from Aristotle. It can be seen most clearly in Aquinas's prologue; Thomas Aquinas, *In Aristotelis Libros Posteriorum Analyticorum Expositio,* edited by Raymund M. Spiazzi (Turin: Marietti, 1955), n.4: "Secunda vero operatio intellectus est compositio vel divisio intellectus, in qua est iam verum vel falsum. Et huic rationis actui deservit doctrina, quam tradit Aristoteles in libro *Peri hermeneias.*" I am not aware of Ockham's having read Aquinas's commentary as he makes reference to it neither in the prologue to the *Commentary* on the *Sentences* nor in the relevant section of *SL* (on demonstration). However, these two acts of the intellect closely mirror those that Aquinas gives in his *Expositio.* The division was common and repeated by many commentators on *Posterior Analytics* (Aegidius Romanus, e.g.). The fact that the division occurs here sheds further light on the relation between the theory of science and the role of *notitia intuitiva* found in the *Ordinatio* and that of *Posterior Analytics.* To the best of my knowledge, no commentator on Ockham has pointed out the similarities between Ockham's division, as presented here, and the division that was common in prologues to commentaries on *Posterior Analytics.* We will turn to this question in detail below.

34. *I Sent.*, 17: " . . . actus iudicativus respectu alicuius complexi praesupponit actum apprehensivum respectu eiusdem."

35. *I Sent.*, 21.

36. *I Sent.*, 21: " . . . intellectus nullam propositionem potest formare, nec per consequens apprehendere, nisi primo intelligat singularia, id est incomplexa."

37. *I Sent.*, 22: " . . . nullus actus partis sensitivae est causa immediata proxima, nec partialis nec totalis, alicuius actus iudicativi ipsius intellectus." Ockham's argument for this relies, at least partially, on the famous principle of economy. He argues that if an intellective, apprehensive act suffices for the judicative act, then it is unnecessary to posit other causes.

38. *I Sent.*, 27: " . . . tales veritates contingentes non possunt sciri de istis sensibilibus nisi quando sunt sub sensu, quia notitia intuitiva intellectiva istorum sensibilium pro statu isto non potest haberi sine notitia intuitiva sensitiva eorum."

39. See chapter 3 above.

40. As noted above, the distinction of acts or operations of the intellect that Aquinas raises in his *prooemium* became standard in commentaries on *Posterior Analytics* throughout the medieval period. Commentators between Aquinas and Ockham (e.g., Aegidius Romanus) as well as commentators

after Ockham were well aware of this distinction and used it in their commentaries. There is scant evidence that Ockham was aware of others' commentaries on *Posterior Analytics*. He shows an awareness of Grosseteste's commentary, and Grosseteste does seem to imply a distinction between *apprehensive* acts and *intellectual* acts, but not *judicative*. Ockham also shows an awareness of the commentary of Walter Burley, which does not use this division (probably because it does not contain a *prooemium*). We do know that Ockham seemed to have thought that he would produce his own *Commentary* on the *Posterior Analytics*, for he mentions such a commentary in *SL*, p. 584. We do not possess such a commentary and it seems most likely that one was not produced. However, the language that Ockham here uses comes only from the commentary tradition on *Posterior Analytics*. It seems unlikely that Ockham would have drawn on that tradition and yet is not aware of it. Ockham must consider *notitia intuitiva* as, if not *intellectus* itself, then at least the beginning of the process of *intellectus*. Furthermore, one cannot ignore the fact that Ockham raises here precisely the same example that one finds in Aquinas's discussion of *intellectus:* an herb that is capable of curing a disease. This example was developed precisely in connection with *intellectus* and the fact that Ockham uses this example shows his awareness that he is addressing himself to the problem of *intellectus*.

41. Notice here that while Aquinas also argues that *intellectus* of God and deity is, in principle, impossible because it "exceeds the grasp of rational comprehension" (*ST* I, 1, 1c), for later thinkers, and especially Scotus, *intellectus* is not impossible *in principle* but only because if we have *intellectus* of God and deity, there would be nothing to hope for in beatitude. In other words, there is nothing about the beatific vision or about the human intellect, according to Scotus, Ockham, and many who come after them (Pierre d'Ailly and Marsilius of Inghen are but two examples), that prohibits *intellectus* of God under the aspect of deity. It is denied to us only because such a grasp of God under deity would leave nothing left for beatitude. In that case, the entire concept of salvation would be a moot point.

42. Indeed, I have yet to find a single modern commentator who places Ockham's discussion of intuitive knowledge within this framework.

43. *Post. Anal.,* n. 6: "Pars autem Logicae, quae primo deservit processui, pars iudicativa dicitur, eo quod iudicium est cum certitudine scientiae. Et quia iudicium certum de effectibus haberi non potest nisi resolvendo in prima principia, ideo pars haec Analytica vocatur, idest resolutoria."

44. *I Sent.,* 17: "Et ille actus apprehensivus potest esse sine iudicativo et non e converso; igitur est prior naturaliter, et ita alius praesupponit ipsum."

45. That error for Aquinas is a logical problem is clear also from his discussion in the first article of *ST:* "quia veritas de Deo per rationem investigata [as opposed to those truths which God reveals to us], a paucis, et per longum tempus, et cum admixtione multorum errorum homini proveniret. . . . Ut igitur salus hominibus et convenientius, et certius proveniat, necessarium fuit, quod de divinis per divinam revelationem instruerentur." Here the

contrast is between those who grasp the truth immediately and those who investigate through reason. The problem with investigating divine truths through reason is that error can enter. Thus, logic must be open to error in a way that the grasping of principles is not.

46. On this process, cf. T. K. Scott, "Ockham on Evidence, Necessity and Intuition," *Journal of the History of Philosophy* 9 (1971): 15–41.

47. This point is made quite nicely in Rudi A. Te Velde, *Participation and Substantiality in Thomas Aquinas* (Leiden: E. J. Brill, 1995), pp. 87–91. There is a problem with the notion of *esse* in Aquinas that, to my knowledge, is pointed out for the first time in this text.

48. When Ockham says that the apprehensive act may also pertain to propositions and demonstrations, he means that through this act I grasp the *proposition* 'Socrates is white' and not that I grasp the *fact* that Socrates is white.

49. *I Sent.*, 22.

50. *I Sent.*, 24: "Et tamen manifestum, est quod de eisdem potest haberi notitia incomplexa, et tamen veritas illa ignorari; ergo respectu illorum terminorum est duplex notitia incomplexa specie distincta." This whole argument is given in a section that is supposed to prove the first conclusion, viz., that the judicative act with respect to some proposition presupposes an apprehensive act with respect to the same proposition. It is clear that Ockham is here including abstractive and intuitive cognition under the heading of apprehensive acts of the intellect.

51. *I Sent.*, 32.

52. On this issue, Leff seems to be on the right track. "Ockham's account of intuitive and abstractive knowledge in the Prologue of the *Ordinatio* shifts the emphasis, compared with Duns, from the contrast between perceptual and conceptual knowledge to that between evidential and inevidential," Leff, *William of Ockham: The Metamorphosis of Scholastic Discourse*, 14. This corrects Leff's earlier treatment of the distinction, which follows most interpreters, as bearing on existential status. "Abstractive knowledge—as apprehension of the terms—is the condition of propositions at all; but intuitive knowledge—as the apprehension of things as well—is the condition and the guarantee of their truth. One is nonexistential, the other is existential," 7. This is flawed in two respects. First, as Leff himself hints, intuitive knowledge can be of terms as well as things. Therefore, it is not clear that Ockham would agree that abstractive knowledge is the requirement of propositions at all. It seems entirely possible that a proposition can be formed that is not based on any abstractive knowledge at all, but only on intuitive. Secondly, to frame the distinction in terms of existence, as many interpreters do, misses Ockham's main point: What are the conditions that make an evident judgment possible? Insofar as the judgment concerns a proposition such as 'X exists', intuitive knowledge concerns existence. Insofar as the judgment concerns a proposition such as 'X is white', it does not concern existence, except as existence is the condition of all

things, and, as such, is convertible with the term "thing" (cf. *SL,* 554). We will return to this theme below.

53. *I Sent.,* 27.

54. *I Sent.,* 27.

55. *I Sent.,* 31: " . . . notitia intuitiva est talis notitia virtute cuius potest sciri utrum res sit vel non. . . ." Here, I have refrained from translating "utrum res sit vel non" as "whether the thing exists or does not exist." The reason for this will become obvious in a moment. However, it should be stated that this phrase does not have to be taken as having purely existential import. Rather, it could be understood as a quasi-formalized expression such as, "whether the X is a Y." This finds support in Ockham's treatment of the copula in the *SL,* 553–55. In that text, Ockham goes to great pains to show that "existence" is not a real predicate.

56. *I Sent.,* 31: " . . . ita quod si res sit, statim intellectus iudicat eam esse et evidenter cognoscit eam esse, nisi forte impediatur propter imperfection illius notitiae."

57. *I Sent.,* 31: "Similiter, notitia intuitiva est talis quod quando aliquae res cognoscuntur quarum una inhaeret alteri vel una distat loco ab altera vel alio modo se habet ad alteram, statim virtute illius notitiae incomplexae illarum rerum scitur si res inhaeret vel non inhaeret, si distat vel non distat, et sic de aliis veritatibus contingentibus."

58. While in the discussion in this question of the prologue Ockham does not explicitly tie the discussion of intuitive knowledge to the idea of certainty, he does so later, *I Sent.,* 200. On the relation between certainty, vision, and intuitive knowledge, see the excellent treatment in Tachau, *Vision and Certitude in the Age of Ockham: Optics, Epistemology and the Foundations of Semantics 1250–1345.*

59. Tachau, ibid., 148.

60. Scott, "Ockham on Evidence, Necessity and Intuition," 46.

61. *I Sent.,* 31.

62. *I Sent.,* 5–6.

63. This charge is made by Adams in "Intuitive Cognition, Certainty and Skepticism in William Ockham," 389–98.

64. I have argued against this position in Richard A. Lee, "Peter Aureoli as Critic of Aquinas on the Subalternate Character of the Science of Theology," *Franciscan Studies* 55 (1998): 121–36; and Richard Lee, "Peirce's Retrieval of Scotistic Realism," *American Catholic Philosophical Quarterly* 72, no. 2 (1998): 179–96. The particular form of skepticism with which Ockham and his followers are charged is not of the ancient sort (which is an ethical prohibition against bothering with the world) but of the modern sort. The charge is further contested by Dallas G. Denery II, "The Appearance of Reality: Peter Aureol and the Experience of Perceptual Error," *Franciscan Studies* 55 (1998): 52: "It is tempting to interpret Aureol's distinction between apparent and real being in terms of more recent distinctions, such as that between subject and object or between phenomena and noumena. But

such interpretations do a disservice to Aureol's text because Aureol did not distinguish them in that manner. . . . For Aureol, the world made itself known, announced itself, through its appearance to the senses and to the intellect and it was precisely this claim, a claim which arose out of assumptions at the very heart of early fourteenth century theories of cognition and perception, which Aureol's successors found so difficult to accept and to ignore." What Denery says here about Aureol applies equally to Ockham. The "announcing itself" of the world is precisely what I have been referring to by "existing singular," which is without ground in reason.

65. This is in distinction to Aquinas's theory of knowledge where, to a certain extent, the knower and the known are one.

66. This is to say nothing about the role of the will in the formation of concepts themselves. I leave this question aside because of Ockham's apparent change of mind on this position.

67. Stephen D. Dumont, "Theology as a Science and Duns Scotus's Distinction Between Intuitive and Abstractive Cognition," *Speculum* 64 (1989): 579–99.

Chapter 6

1. William of Ockham, *Scriptum in Librum Primum Sententiarum Ordinatio,* edited by Gedeon Gál, O.F.M. and Stephen Brown, O.F.M., Guillelmi de Ockham Opera Philosophica et Theologica, OT I (St. Bonaventure: Franciscan Institute Publications, 1967), p. 115. Hereafter cited as *I Sent.,* followed by page number.

2. *I Sent.,* 5: " . . . intellectus viator est ille qui non habet notitiam intuitivam deitatis sibi possiblem de potentia dei ordinata. Per primum excluditur intellectus beati, qui notitiam intuitivam dei habet; per secundum excluditur intellectus damnati, cui non est illa notitia possibilis de potentia dei ordinata, quamvis sit sibi possibilis de potentia dei absoluta." We will turn to the distinction between the "absolute" and "ordered" power of God below.

3. *I Sent.,* 118.

4. See the definition of the intellect of the *viator* given in note 2 above.

5. *I Sent.,* 189: " . . . habitus principiorum est notitior et evidentior habit conclusionum, ergo impossible est quod principia tantum credantur et conclusiones sciantur."

6. *I Sent.,* 206: " . . . qoud amnis habitus veridicus evidens respectu veritatis necessariae est sapientia vel scientia etc.; tamen habitus veridicus inevidens potest esse fides, et talis est theologia pro magna sui parte. Similiter, respectu talis veritatis est aliquis habitus qui non est proprie veridicus, quia non est iudicativus sed tantum apprehensivus, et talis est theologia pro aliqua sui parte."

7. William of Ockham, *Quodlibeta Septem,* edited by J. C. Wey, Guillelmi de Ockham Opera Philosophica et Theologica, OT IX (St. Bonaventure: Franciscan Institute Publications, 1980), p. 586, hereafter cited as *Quod.,*

followed by page number. This is the most lucid account of the distinction between God's absolute power (*potentia dei absoluta*) and ordered power (*potentia dei ordinata*) that Ockham puts to work throughout his corpus. On this distinction, a general consensus seems to have emerged around Courtenay's thesis that the absolute power of God is understood by Ockham to mean all the possibilities initially open to God rather than God changing or undoing what has been done. William Courtenay, "Nominalism and Late Medieval Religion," in *The Pursuit of Holiness in Late Medieval and Renaissance Religion,* edited by Charles Trinkhaus and Heiko Oberman (Leiden: E. J. Brill, 1974), pp. 26–59. For a history of the distinction in medieval thought, see William Courtenay, *Capacity and Volition: A History of the Distinction of Absolute and Ordained Power* (Bergamo: P. Lubrina, 1990); and Francis Oakley, *Omnipotence, Covenant and Order. An Excursion in the History of Ideas from Abelard to Leibniz* (Ithaca: Cornell University Press, 1984). Other important works on the topic include: M. A. Pernoud, "The Theory of the Potentia Dei According to Aquinas, Scotus and Ockham," *Antonianum* 47 (1972): 69–95; Jürgen Miethke, *Ockhams Weg Zur Social Philosophie* (Berlin: Walter de Gruyter, 1969), pp. 137–56; and Heiko Oberman, "*Via Antiqua* and *Via Moderna*: Late Medieval Prolegomena to Early Reformation Thought" in *From Ockham to Wyclif,* edited by Anne Hudson and Michael Wilks (Oxford: Basil Blackwell, 1987), pp. 445–63.

8. I have in mind here scholars such as Gilson, Grabmann, De Wulf, and to a certain extent Michalski. The partisan lines in this dispute have been clearly drawn. Courtenay has effectively shown that the interpretation by later commentators of the distinction of God's power has a decidedly religious aspect, even when the interpreters do not approach the material from a confessional standpoint. That is to say, the terms of the debate are still largely built around the categories of the hermeneutic tradition. See Courtenay, "Nominalism and Late Medieval Religion," p. 32; Courtenay, *Capacity and Volition: A History of the Distinction of Absolute and Ordained Power,* pp. 11–24.

9. Courtenay, "Nominalism and Late Medieval Religion," p. 39: "*Potentia absoluta* and *potentia ordinata* are not, therefore, two ways in which God can or might act, normally and with the concurrence of nature in the case of the latter and extraordinarily, supernaturally, and miraculously in the case of the former. *Potentia absoluta* referred to the total possibilities *initially* open to God, some of which were realized by creating the established order; the unrealized possibilities are now only hypothetically possible." See also, Courtenay, *Capacity and Volition: A History of the Distinction of Absolute and Ordained Power,* p. 120. "*Potentia absoluta* is not a sphere of action but of possibility." Miethke argues basically the same point: " . . . wie bei Duns ist bei Ockham die potentia absoluta Grund und Bedingung der faktischen Heilsordnung de potentia ordinata. . . . Die Redeweise 'de potentia absoluta' fragt nur nach dem, was überhaupt geschehen kann, die Redeweise 'de potentia ordinata' geht aus von der faktischen gesetzten Ordnung . . . was Gott de potentia ordinata tut, ein

in das weite Feld von Gottes Möglichkeiten." Miethke, *Ockhams Weg zur Social Philosophie*, p. 152.

10. *Quod.*, 586: "Nec sic est intelligenda quod aliqua potest ordinate facere, et aliqua potest absolute et non ordinate, quia Deus nihil potest facere inordinate." Ockham concludes the same thing in William of Ockham, *Summa Logicae*, edited by Philotheus Boehner, O.F.M., Gedeon Gál, O.F.M., and Stephan Brown, Guillemi de Ockham Opera Philosophica et Theologica, OP I (St. Bonaventure: Franciscan Institute Publications, 1974), pp. 779–80; hereafter cited as *SL*, followed by page number.

11. Courtenay, "Nominalism and Late Medieval Religion," p. 46.

12. That is, in my opinion, how the "great synthesis" can be achieved at all. It is, in a sense, a ruse that can be perpetrated only if one ignores either faith or reason. Furthermore, it is not even something that Aquinas promises, but rather is a modern fantasy. Ockham, in that sense, achieved a greater synthesis than Aquinas because he allowed each (faith and reason) its own sphere and thus made them finally compatible.

13. *I Sent.*, 180.

14. See chapter 5 above.

15. William of Ockham, *Scriptum in Librum Primum Sententiarum Ordinatio, Distinctiones IV–XVIII*, edited by Girard Etzkorn, Guillelmi de Ockham Opera Philosophica et Theologica, vol. OT III (St. Bonaventure: Franciscan Institute Publications, 1970), pp. 355ff.

16. Ibid., 354–55: "Videtur tamen quod evidentius potest probari primitas efficientis per conservationem rei a sua causa quam per productionem. . . ."

17. See his response to various arguments in favor of a scientific habit of theology, *I Sent.*, 190–91.

18. *I Sent.*, 76.

19. On this question, see my, "Contingency from Necessity: The Late Medieval Grounding of Scientific Necessity," Ph.D. Diss. (Krakow, Poland: Jagiellonian University, 1995).

20. Jürgen Miethke, op. cit., 158.

21. According to Ockham's theory of supposition, the proposition 'All humans are animals', the term "human" has confused, distributive supposition. The proposition can be analyzed as "this human is an animal and that human is an animal, etc." But the several subjects ("this human, that human") are verified of actually existing humans. The proposition only becomes necessary if it is taken conditionally or hypothetically. See *SL*, 210–11.

22. "Der theologische Absolutismus verweigert dem Menschen den Einblick in die Rationalität der Schöpfung. . . ." Hans Blumenberg, *Die Legitimität der Neuzeit*, 2nd (Frankfurt am Main: Suhrkamp, 1988), p. 164; Hans Blumenberg, *The Legitimacy of the Modern Age*, translated by Robert M. Wallace (Cambridge, Mass.: MIT Press, 1983), p. 149.

23. *SL*, 554.

24. Ibid., 250: " . . . [ad veritatem] sufficit et requiritur quod subiectum et praedicatum supponant pro eodem."

25. *I Sent.*, 31.
26. *I Sent.*, 31. What kind of imperfection there might be in this knowledge, Ockham does not specify. If there is some kind of imperfection, however, might this not indicate that such knowledge is not as immediate as Ockham states it is?
27. *Metaphysics* 1011b25; Aristotle, *Metaphysics,* translated by Hippocrates G. Apostle (Grinnell, Iowa: The Peripatetic Press, 1979), p. 70.
28. Anselm, *De Veritate,* vol. I of *S. Anselmi Cantauriensis Archiepiscopi Opera Omnia,* edited by Francis Schmitt, O.S.B. (Edinburgh: Thomas Nelson and Sons, 1946), p. 178: "M. Quid igitur tibi videtur ibi veritas? D. Nihil aliud scio nisi quia cum significat esse quod est, tunc est in eas veritas et est vera."
29. Anselm, like many thinkers in an Augustinian tradition, maintains a distinction between propositional truth, which is defined according to correctness, and a more ontological notion of truth that belongs to a thing by virtue of its being like or unlike a divine idea.
30. Aristotle, *Metaphysica,* edited by Gudrun Vuillemin-Diem, translated by Guillelmus de Moerbeka, Aristoteles Latinus, vol. XXV 3.2 (Leiden: E. J. Brill, 1995), p. 89: "Dicere namque ens non esse aut HOC esse falsum, ens AUTEM esse et non ens non esse uerum; quare ET DICENS esse aut non uerum dicet aut mentietur; sed neque ens dicit non esse aut esse neque non ens."
31. To be more precise, 'Socrates is white' is true if both "Socrates" and "white" stand for the same thing—the thing Socrates and the thing in which whiteness inheres must be one and the same thing.
32. This seems implied by Ockham's discussion in the *SL.* See note 23 above.
33. *Metaphysics,* 1011b20; the translation is from Apostle, op. cit., p. 70.
34. The entire issue can be phrased in Kant's terms: Is existence a real predicate?
35. *I Sent.*, 38.
36. Adding a negation, the same could be said of nonexisting things as well.
37. *Quod.,* 498: "dico quod Deus non potest causare in nobis cognitionem talem per quam evidenter apparet nobis rem esse praesentem quando est absens, quia includit contradictionem."
38. William of Ockham, *Expositio in Librum Praedicamentorum Aristotelis,* edited by Gedeon Gál, Guillelmi de Ockham Opera Philosophica et Theologica, vol. OP II (St. Bonaventure: Franciscan Institute Publications, 1978), pp. 238ff.
39. *Quod.,* 499.
40. *I Sent.*, 39: " . . . Deus habet notitiam intuitivam omnium, sive sint sive non sint, quia ita evidenter cognoscit creaturas non esse quando non sunt, sicut cognoscit eas esse quando sunt."
41. It could also be said that it is not an ontological or even a metaphysical issue for Ockham. Existence, for Ockham, is nothing other than "thing." "Thing" is the basic presupposition that any metaphysics and certainly any epistemology must presuppose. Ockham's theory of intuitive knowledge, then, shows just that presupposition.

42. Blumenberg, *The Legitimacy of the Modern Age,* p. 151; Blumenberg, *Die Legitimität der Neuzeit,* p. 167.

Chapter 7

1. Marsilius was born around 1340 and died in 1396. D'Ailly was born in 1351 and died in 1420. I will give brief biographical details for each below.
2. For this biographical information, I draw on M. J. F. M. Hoenen, *Marsilius of Inghen: Divine Knowledge in Late Medieval Thought* (Leiden: E. J. Brill, 1993), pp. 7–11, as well as on the introduction to Marsilius of Inghen, *Quaestiones super Quattuor Libros Sententiarum,* edited by Manuel Santos Noya (Leiden: E. J. Brill, 2000), pp. xvii-xvi.
3. Hoenen, ibid., points out the anachronism involved in counting Marsilius among the *moderni,* and even in referring to his thought as nominalist: "The view that Marsilius was a nominalist occurs in much of the modern literature. It is based essentially on the fifteenth- and sixteenth-century sources discussed above. The nominalism found there is then projected back into the fourteenth century." Hoenen points out that not only is this an anachronism, but the views of thinkers labeled "nominalist" were in fact too varied to be included in one school.
4. See Hoenen, op. cit., 19.
5. Marsilius of Inghen, *Quaestiones super Quattuor Libros Sententiarum,* 61.
6. Ibid., 65.
7. Ibid.: "Et dividitur prima divisione in notitiam, quae est causa rerum, et a nullo causata—quae est notitia divina . . . et in notitiam causatam a rebus vel a re et est notitia in cognoscentibus dependentibus."
8. Ibid.: "Incomplexa est simplex singularis apprehensio rei, ut visus exterior apprehendit coloratum una simplici apprehensione singulari representante rem cum omnibus suis proprietatibus accidentalibus a visu perceptibilibus. . . ."
9. Due to the similarity of this way of describing simple, sensitive cognition, I take it that this is *"notitia intuitiva."*
10. Ibid., 66: "Complexa est, ut cum sensus dicit hoc visum esse hoc dulce, etc. Et quod hoc possit sensus, patet de sensu communi comparante diversa sensibilia propria diversorum sensuum exteriorum. . . ."
11. Ibid., 68: "Notitiarum intellectivarum aliqua est complexa, aliqua est incomplexa. . . . Incomplexarum aliqua est singularis et aliqua communis."
12. Ibid.: "Singularis est duplex, scilicet vaga, quae sequitur notitiam sensus; et determinata, quae dificillima est inter notitias incomplexas, quia distinctissima. Et est illa cui correspondent individua predicamenti substantiae, sicut conceptus, quibus correspondent hi termini 'Socrates' et 'Plato'. Communis est simplex apprehensio rei ex modo suae significationis communis multis suppositis."
13. Ibid., 69: " . . . non sint res universales in essendo, tamen res singulars aliquae sunt essentialiter similiores qaum aliae. Et ab ipsis similibus per ab-

stractionem intellectus, quam facit naturaliter et non libere, sumuntur conceptus universales iuxta ordinem conceptuum singularium vagorum secundum maiorem distinctionem et minorem. . . ."

14. Ibid., 69: "Sexta divisio: Complexarum notitiarum aliqua est complexa complexione propositionali, quae vocatur a quibusdam distans; aliqua est complexa complexione non propositionali, quae vocatur indistans, ut complexio qua dicit anima 'homo albus', etc."

15. Ibid.: "Septima divisio: Notitiarum propositionalium aliqua est apprehensiva, et est qua apprehenditur sensus quem propositio iuxta significationem suorum terminorum importat. . . ."

16. As we saw, Ockham maintains that there is such a kind of intuitive knowledge, though it is based on intuitive knowledge of the terms of the proposition.

17. Ibid.: " . . . alia est assensiva, qua propositioni apprehensae assentimus. Ut discipulus audiens conclusionem quam magister vult probare, ante probationem habet de ea notitiam apprehensivam, quia intelligit quod significet, sed non habet assensivam, donec probatur. Est autem apprehensiva ipsamet propositio intellecta ut aestimo; sed adhaesiva vel assensiva videtur esse distincta a propositione, eo quod propositio est, antequam intellectus sibi assentit, et manet in intellectu, postquam eius assensus est corruptus."

18. Ibid., 70: "Octava divisio: Assensivarum notitiarum alia est sine probatione praevia, alia est cum probatione praevia. Sine probatione est quadruplex: Quaedem enim singularium, quae vocatur notitia assensiva sensus vel sequens sensum, ut quod hoc tactum sit calidum vel quod aliquid est calidum. Secunda est quae sequitur experientiam cum iuvamine intellectus, ut quod omnis ignis est calidus. Tertia est principiorum evidentium ex implicatione suorum terminorum, ut quod totum est maius sua parte. Quarta est assensus genitus ex auctoritate dicentis; sicut credimus quod Deus est trinus et unus vel quod in ecclesia dicitur missa, aliquo nobis hoc dicente. Prima harum vocatur sensus vel sequens sensum. Secunda vocatur intellectus vel certitudo, quia ex natura intllectus est nota, similiter et tertia. Et quarta vocatur fides."

19. There is one further division, but it is not really a division. The ninth division simply states that any intellective knowledge can be either actual or habitual. This does not give a principle of division, but rather describes two ways in which such knowledge could exist in the soul.

20. I understand the question I am pursuing here to be the question that traditionally fell under the notion of "evidence," even though Marsilius does not use that term in this context.

21. This would depend on what the "terms" are. For if the proposition is something like "that white man is running," then nonpropositionally complex, intellective knowledge would be required as a mediating step.

22. For biographical information, see Laura Ackerman Smoller, *History, Prophecy, and the Stars: The Christian Astrology of Pierre d'Ailly 1350–1420* (Princeton: Princeton University Press, 1994); and Olaf Pluta, *Die Philosophische Psychologie*

Des Peter von Ailly: Ein Beitrag Zur Geschichte der Philosophie Des Späten Mittelalters, Bochumer Studien Zur Philosophie 6 (Amsterdam: B. R. Grüner, 1987). They each refer to other biographical sources that may profitably be consulted.
23. We can point, in particular, to Prol. q. 1, a. 3, which has entire sections that are taken verbatim from Ockham's text. The main exception is that d'Ailly has many arguments against Gregory of Rimini, which obviously do not come from Ockham. With these few exceptions, however, the texts are remarkably similar.
24. "Unde dico quod duplex est evidentia, quaedam est evidentia absoluta qualis est evidentia primi principii vel reducibilis ad eam, alia est evidentia condicionata qualis est evidentia nostri ingenii qua est citra primam." *Petrus de Alliaco, Quaestiones super Libros Sententiarum* (Frankfurt: Minerva, 1968), I, 1, E. All citations are from the prologue. Question number, article number, followed by letter will be given.
25. Ibid.
26. Ibid.: " . . . omnis evidentia est assensus licet non e contra."
27. Ibid.: "Et dicitur verus ad differentia erroris seu assensus falsi vel erronei quantumcunquam sit firmus tamen nunquam est evidens. Secunda pars est sine formidine ad differentia opinionis suspicionis vel coniecture qua non sunt sine formidine de facto vel de possibili. Tertia est causatus naturaliter id est ex causis necessitantibus intellectum ad sic assentiendum ad differentiam fidei qua licet sit assensus sine formidine, tamen non causatus naturaliter sed libere, etc. Quarta est quo non est possible, etc. Ad differentiam assensus causati per sillogismum falsigraphum quo licet possit habere condiciones praedictas, tamen habens illum assensum sic assentiendo decipitur, non per assensum conclusum sed per medium conclusivum. Item pars ponitur ad differentia evidentiae secundum quid sive condicionate."
28. Ibid. The entire passage reads: "Evidentia autem secundum quid potest describi quod est assensus verus sine formidine causatus naturaliter quo non est possibile stante dei influentia generali et nullo facto miraculo intellectum assentire et in sic assentiendo decipi vel errare."
29. I, 1, F: "Tertia conclusio est quod impossibile est viatorem aliquid extrinsecum ab eo sensibile evidenter cognoscere esse evidentia simpliciter absoluta sed bene evidentia secundum quid et condicionata."
30. Ibid.: " . . . loquendo de evidentia secundum quid seu condicionata vel ex suppositione scilicet stante dei influenti generali et cursu nature solito nullo quam facto miraculo talia possunt esse nobis sufficienter evidentia sic quod de ipsis non habemus rationabiliter dubitare."
31. I, 1, H: " . . . aliquis posset dissentire primo principio non sequitur quod illud non sit evidens, sed solum quod potest non esse evidens . . . aliquis dissentiendo primo principio vel assentire eius opposito erraret in fide, tamen quod aliquis possit dissentire primo principio vel assentire eius opposito non contradicit fidei nostre. Primum patet, quia ex opposito primi principii sequitur oppositum cuiuslibet articuli fidei."

32. It might be noted that Leonard A. Kennedy, *Peter of Ailly and the Harvest of Fourteenth Century Philosophy* (Lewiston: Edwin Mellen Press, 1986), p. 15, argues that according to d'Ailly, one can dissent from the first principle because God can cause this dissent by his omnipotence. Yet the text Kennedy cites comes within an objection *against* a position that d'Ailly had asserted. Kennedy never indicates this. It is true that in his response to the argument, d'Ailly seems to accept the position: "Secundum patet quia dicere quod deus possit huiusmodi assensum vel dissensum causare non contradicit fidei, immo videtur favere articulo de omnipotentia dei." This is a strange response. For d'Ailly neither completely accepts nor completely rejects the possibility that I could dissent from the principle of noncontradiction. His only response is that it does not contradict the faith, but rather supports the belief in the omnipotence of God. It must be noted, however, that even if God causes dissent to this principle, the falsity of the principle does not follow.

Postlude

1. While Heidegger speaks of "metaphysics" in many places, perhaps the best source of his thought on this issue, particularly in relation to its end and completion are three related texts written in the 1930s: Martin Heidegger, *Beiträge Zur Philosophie (Vom Ereignis)*, edited by Friedrich-Wilhelm von Hermann, Gesamtausgabe, 65 (Frankfurt: Vittorio Klostermann, 1989); *Besinnung,* edited by Friedrich-Wilhelm von Hermann, Gesamtausgabe, 66 (1997); Heidegger, *Die Geschichte des Seyns,* edited by Peter Trawny, Gesamtausgabe, 69 (Frankfurt: Vittorio Klostermann, 1998).

2. Martin Heidegger, *Der Satz Vom Grund* (Pfullingen: Verlag Gunther Neske, 1957); *Principle of Reason,* translated by Reginald Lilly (Bloomington: Indiana University Press, 1991).

3. The givenness of the universality of thought is not peculiar to Aristotelian and medieval thought. The notion that conceptual thought is universal is one of the main features of modern, subjectivist philosophy as well. I will return to this point below in relation to Adorno's thought.

4. This gesture of philosophizing the excess operates both in Heidegger's notion of *Abgrund* as the event of the coming to presence of rational ground and his move to attempt to think in relation to the eventing character [*Ereignis*] of being and in Derrida's attempt to think a notion of difference [*différance*] that is attentive to the fact that difference as such cannot function as a foundational identity. For Heidegger's notion of *Ereignis*, see Heidegger, *Beiträge zur Philosophie (Vom Ereignis).* For Derrida's notion of differance, see Jacques Derrida, *Speech and Phenomena and Other Essays on Husserl's Theory of Signs,* translated by David B. Allison (Evanston: Northwestern University Press, 1973), pp. 129–60.

5. In this regard, see the first section of *Negative Dialectics;* Theodor W. Adorno, *Negative Dialektik,* edited by Rolf Tiedmann, Gesammelte Schriften, 6 (Frankfurt: Suhrkamp, 1997), pp. 7–412; Theodor W. Adorno,

Negative Dialectics, translated by E. B. Ashton (New York: Continuum, 1992), as well as his essay "On Subject and Object," which appears as a "dialectical epilogue" to Theodor W. Adorno, *Stichworte,* edited by Rolf Tiedmann, Gesammelte Schriften, 10/2 (Frankfurt: Suhrkamp, 1997), pp. 595–782.

BIBLIOGRAPHY

Adam Wodeham. *Lectura Secunda.* Edited by Rega Wood. St. Bonaventure: Franciscan Institute Publications, 1990.

Adams, Marilyn. Intuitive Cognition, Certainty and Skepticism in William Ockham. *Traditio* 26 (1970): 389–98.

———. What Does Ockham Mean by Supposition? *Notre Dame Journal of Formal Logic* 12 (1976): 372–91.

———. Ockham's Nominalism and Unreal Entities. *Philosophical Review* 86 (1977): 144–76.

———. Universals in the Early Fourteenth Century. In *The Cambridge History of Later Medieval Philosophy,* edited by Norman Kretzman, Anthony Kenny, and Jan Pinborg, 411–39. Cambridge: Cambridge University Press, 1982.

———. *William Ockham.* Notre Dame: University of Notre Dame Press, 1987.

Adorno, Theodor W. *Negative Dialectics.* Translated by E. B. Ashton. New York: Continuum, 1992.

———. *Negative Dialektik.* Edited by Rolf Tiedmann, 7–412. Gesammelte Schriften 6. Frankfurt: Suhrkamp, 1997.

———. *Stichworte.* Edited by Rolf Tiedmann, 595–782. Gesammelte Schriften 10.2. Frankfurt: Suhrkamp, 1997.

Aegidius Romanus. *Super Libros Posteriorum Analyticorum.* Venice, 1488.

———. *In Petri Lombardi Sententiarum Librum Primum Commentum.* Venice, 1492.

———. *Exposition in Artem Veterem.* Venice, 1507.

Allen, Don Cameron. *Doubt's Boundless Sea: Skepticism and Faith in the Renaissance.* Baltimore: Johns Hopkins Press, 1964.

Analytica Posteriora. Edited by L. Minio-Paluello and B. Dod. Aristoteles Latinus 4. Paris: Desclée de Brouwer, 1968.

Anselm. *De Veritate.* Vol. I of *S. Anselmi Cantauriensis Archiepiscopi Opera Omnia.* Edited by Francis Schmitt, O.S.B., 169–99. Edinburgh: Thomas Nelson and Sons, 1946.

Antweiler, A. *Der Begriff der Wissenschaft bei Aristoteles.* Bonn: P. Hanstein, 1936.

Apollinaris Cremonensis. *Expositio in Primum Posteriorum Aristotelis.* Venice: O de Luna, 1497.

Ariew, Roger. Christopher Clavius and the Classification of Sciences. *Synthese* 83 (1990): 293–300.

Aristotle. *De Caelo.* Translated by W. K. C. Guthrie. Loeb Classical Library 338. Cambridge: Harvard University Press, 1939.

————. *Analytica Posteriora*. Edited by Laurentius Minio-Paluello and Bernard G. Dod. Aristoteles Latinus IV. Brussels: Desclée de Brouwer, 1968.

————. *Posterior Analytics*. Translated by Jonathan Barnes. Oxford: Clarendon Press, 1975.

————. *Metaphysics*. Translated by Hippocrates G. Apostle. Grinnell, Iowa: The Peripatetic Press, 1979.

————. *Categories*. Translated by Hippocrates G. Apostle. Grinnell, Iowa: The Peripatetic Press, 1980.

————. *Physics*. Translated by Hippocrates G. Apostle. Grinnell, Iowa: The Peripatetic Press, 1980.

————. *Posterior Analytics*. Translated by Hippocrates G. Apostle. Grinnell, Iowa: The Peripatetic Press, 1981.

Ashworth, E. J. Can I Speak More Clearly Than I Understand? A Problem of Religious Language in Henry of Ghent, Duns Scotus and Ockham. *Historiographia Linguistica* 7 (1980): 29–38.

————. Signification and Modes of Signifying in 13th Century Logic: A Preface to Aquinas on Analogy. *Medieval Philosophy and Theology* 1 (1991): 39–67.

————. Analogy and Equivocation in 13th Century Logic: Aquinas in Context. *Medieval Studies* 54 (1992): 94–135.

————. Equivocation and Analogy in 14th Century Logic: Ockham, Burley and Buridan. In *Historia Philosophiae Medii Aevi: Studien zur Geschichte der Philosophie des Mittelalters,* edited by B. Mojsisch and O. Pluta. Amsterdam: B. R. Grüner, 1992.

Bañes, Dominico. *Scholastica Commentaria in Primam Partem Summae Theologicae s. Thomae Aquinatis*. Madrid: Editorial F.C.D.A., 1934.

Barnes, Jonathan. Aristotle's Theory of Demonstration. In *Science,* edited by Jonathan Barnes, Malcolm Schofield, and Richard Sorabji, 65–87. Articles on Aristotle 1. London: Duckworth, 1975.

————. Proof and the Syllogism. In *Aristotle on Science: The Posterior Analytics,* edited by Enrico Berti, 17–59. Padua: Editrice Antenore, 1981.

Becher, Hubert. *Gottesbegriff und Gottesbeweis bei Wilhelm von Ockham. Scholastik* (1928): 369–93.

Beckmann, Jan P. "Scientia Proprie Dicta": Zur Wissenschaftstheoretischen Grundlegung der Philosophie bei Wilhelm von Ockham. In *Sprache und Erkenntnis im Mittelalter,* edited by Jan P. Beckmann et al., 637–47. Miscellanea Mediaevalia 13/2. Berlin: Walter de Gruyter, 1981.

————. Weltkontingenz und Menschliche Vernunft bei Wilhelm Ockham. In *L'Homme et son Univers au Moyen Age,* edited by Christian Wenin. Louvain: Editions de L'Institut Supérieur de Philosophie, 1986.

————. Allmacht, Freiheit und Vernunft. Zur Frage nach "Rationalen Konstanten" im Denken des Späten Mittelalters. In *Philosophie im Mittelalter: Entwicklungslinien und Paradigmen,* edited by J. P. Beckmann et al., 275–93. Hamburg: Felix Meiner Verlag, 1987.

Bettoni, Efrem, O.F.M. *Duns Scotus: The Basic Principles of His Philosophy.* Translated by Bernardine Bonansea, O.F.M. Washington: The Catholic University of America Press, 1961.

Blumenberg, Hans. Licht Als Metapher in der Wahrheit. *Studium Generale* 10 (1957): 432–47.

———. *Die Genesis der Kopernikanischen Welt.* Frankfurt: Suhrkamp, 1981.

———. *The Legitimacy of the Modern Age.* Translated by Robert M. Wallace. Cambridge, Mass.: MIT Press, 1983.

———. *Genesis of the Copernican World.* Translated by Robert Wallace. Cambridge, Mass.: MIT Press, 1987.

———. *Die Legitimität der Neuzeit.* 2nd ed. Frankfurt am Main: Suhrkamp, 1988.

Bochenski, Joseph. Die Fünf Wege. *Freiburger Zeitschrift Für Philosophie und Theologie* 36 (1989): 235–65.

Boehner, Philotheus, O.F.M. The Notitia Intuitiva of Non-Existents According to William of Ockham. *Traditio* 1 (1943): 223–75.

———. Zu Ockhams Beweis der Existenz Gottes. *Franziskanische Studien* 32 (1950): 50–69.

———. *Collected Articles on Ockham.* Edited by E. Buytaert. St. Bonaventure: Franciscan Institute Publications, 1958.

Boler, John F. *Charles Peirce and Scholastic Realism.* Seattle: University of Washington Press, 1963.

———. Scotus and Intuition: Some Remarks. *Monist* 49 (1965): 551–70.

———. Ockham on Intuitive Cognition. *Journal of the History of Philosophy* 11 (1973): 95–106.

———. Ockham on Evident Cognition. *Franciscan Studies* 36 (1976): 85–98.

Bos, E. P. *Medieval Semantics and Metaphysics.* Nijmegen: Ingenium Publishers, 1985.

Brampton, Charles. Scotus, Ockham, and the Theory of Intuitive Cognition. *Antonianum* 40 (1965): 111–42.

Brown, Jerome V. John Duns Scotus on Henry of Ghent's Theory of Knowledge. *The Modern Schoolman* 56 (1978): 1–29.

———. The Meaning of *notitia* in Henry of Ghent. In *Sprache und Erkenntnis im Mittelalter,* edited by J. P. Beckmann et al., 992–98. Miscellanea Mediaevalia 13/2. Berlin: Walter de Gruyter, 1981.

Brown, Stephen. Sources for Ockham's Prologue to the Sentences. *Franciscan Studies* 26 (1966): 36–65.

Buijs, Joseph A. The Negative Theology of Maimonides and Aquinas. *Review of Metaphysics* 41 (1988): 723–38.

Burnyeat, M. F. Aristotle on Understanding Knowledge. In *Aristotle on Science: The Posterior Analytics,* edited by Enrico Berti, 97–139. Padua: Editrice Antenore, 1981.

Burrows, Mark S. Naming the God Beyond Names: Wisdom from the Tradition on the Old Problem of God-Language. *Modern Theology* 9 (1993): 37–53.

Catto, J. L. *Theology and Theologians 1220–1320.* Oxford: Clarendon Press, 1984.

Chartularium Universitatis Parisiensis. Edited by Henricus Denifle, O.P. Paris: Culture et Civilisation, 1964.

Chenu, M. D. *The Scope of the Summa of St. Thomas.* Translated by R. E. Brennan. Washington: Thomist Press, 1958.

Coombs, Jeffrey. Jeronimo Pardo on the Necessity of Scientific Propositions. *Vivarium* 33 (1995): 9–26.

Copernicus, Nicolaus. *On the Revolutions of the Heavenly Spheres.* Translated by Charles Glenn Wallis. Amherst: Prometheus Books, 1995.

Courtenay, William. Nominalism and Late Medieval Religion. In *The Pursuit of Holiness in the Late Medieval and Renaissance Religion,* edited by Heiko Oberman and Charles Trinkhaus, 26–59. Leiden: E. J. Brill, 1975.

————. *Covenant and Causality in Medieval Thought: Studies in Philosophy, Theology and Economic Practice.* London: Varorium Reprints, 1984.

————. The Dialectic of Omnipotence in the High and Late Middle Ages. In *Divine Omniscience and Omnipotence in the Middle Ages,* edited by T. Rudavsky, 243–70. Dordrecht: D. Reidel, 1985.

————. *Capacity and Volition: A History of the Distinction of Absolute and Ordained Power.* Bergamo: P. Lubrina, 1990.

Crombie, A. C. *Robert Grosseteste and the Origins of Experimental Science 1100–1700.* Oxford: Clarendon Press, 1953.

Davis, Leo Donald. The Intuitive Knowledge of Non-Existents and the Problem of Late Medieval Skepticism. *New Scholasticism* 49 (1975): 410–30.

Day, Sebastian. *Intuitive Cognition: A Key to the Significance of the Later Scholastics.* St. Bonaventure: The Franciscan Institute, 1947.

Denery II, Dallas G. The Appearance of Reality: Peter Aureol and the Experience of Perceptual Error. *Franciscan Studies* 55 (1998): 27–52.

Derrida, Jacques. *Speech and Phenomena and Other Essays on Husserl's Theory of Signs.* Translated by David B. Allison. Evanston: Northwestern University Press, 1973.

Desharnais, Richard. Reassessing Nominalism: A Note on the Epistemology and Metaphysics of Pierre d'Ailly. *Franciscan Studies* 34 (1974): 296–305.

Dewan, Lawrence. "Obiectum": Notes on the Invention of a Word. *Archives d'Histoire Doctrinale et Littéraire au Moyen Age* 48 (1981): 37–96.

Dominicus Soto. *In Libros de Demonstratione Commentaria.* Frankfurt: Minerva, 1967.

Dreiling, R. *Der Konzeptualismus Des Petrus Aureoli.* Münster: Aschendorff, 1913.

Duhem, Pierre. *Medieval Cosmology.* Translated by Roger Ariew. Chicago: University of Chicago Press, 1985.

Dumont, Richard. Scotus' Intuition Viewed in the Light of the Intellect's Present State. In *De Doctrina Ionnis Duns Scoti in Studia Scholastico-Scotistica II.* Rome: Societas Internationalis Scotistica, 1968.

Dumont, Stephen D. Theology as a Science and Duns Scotus's Distinction Between Intuitive and Abstractive Cognition. *Speculum* 64 (1989): 579–99.

————. The *Propositio Famosa Scoti:* Duns Scotus and Ockham on the Possibility of a Science of Theology. *Dialogue* 31 (1992): 415–29.

Duns Scotus. *Opera Omnia.* Edited by L. Vives. 26 vols. Paris, 1891–1895.

————. *Opera Omnia.* Edited by Commissio Scotisticae. Vatican City: Typis Polyglottis Vaticanis, 1950-.

————. *Treatise on God as First Principle.* Translated by Allan B. Wolter O.F.M. Chicago: Franciscan Herald Press, 1966.

————. *Quaestiones super Libros Metaphysicorum Aristotelis. Libri I-V.* Edited by R. Andrews et al. Opera Philosophica 3. St. Bonaventure: The Franciscan Institute, 1997.

————. *Quaestiones super Libros Metaphysicorum Aristotelis. Libri VI-IX.* Edited by R. Andrews et al. Opera Philosophica 4. St. Bonaventure: The Franciscan Institute, 1997.

Durandus a Sancto Porciano. *In Petri Lombardi Sententias Theologicas Commentariorum Libri IIII.* Venice, 1571.

Elders, Leo. *Faith and Science: An Introduction to St. Thomas' Expositio in Boethii de Trinitate.* Rome: Herder, 1974.

————. *The Metaphysical Theology of St. Thomas Aquinas.* Leiden: E. J. Brill, 1990.

————. *The Metaphysics of St. Thomas Aquinas in Historical Perspective.* Leiden: E. J. Brill, 1993.

Esser, Thomas, O.P. *Die Lehre des Hl. Thomas von Aquino Über die Möglichkeit einer Anfanglosen Schöpfung.* Münster: Aschendorff, 1895.

Farthing, John L. *Thomas Aquinas and Gabriel Biel: Interpretations of St. Thomas Aquinas in German Nominalism on the Eve of the Reformation.* Duke Monographs in Medieval and Renaissance Studies 9. Durham: Duke University Press, 1988.

Foucault, Michel. *The Order of Things.* New York: Vintage Books, 1973.

Frassen, R. P. Claudius. *Scotus Academicus.* Rome: Sullustiana, 1900.

Gilson, Etienne. The Road to Skepticism. In *The Unity of Philosophical Experience,* 61–91. New York: Charles Scribner's Sons, 1937.

Goddu, André. *The Physics of William of Ockham.* Leiden: E. J. Brill, 1984.

————. The Dialectic of Certitude and Demonstrability According to William of Ockham and the Conceptual Relation of His Account to Later Developments. In *Studies in Medieval Natural Philosophy,* edited by Stefano Caroti, 95–131. Firenze: L. S. Olschki, 1989.

Gracia, Jorge J. E. Thomas on Universals. In *Thomas Aquinas and His Legacy,* edited by David M. Gallagher, 16–36. Washington: The Catholic University of America Press, 1994.

Grant, Edward. *Studies in Medieval Science and Natural Philosophy.* London: Variorum Reprints, 1981.

Guerrière, Daniel. The Aristotelian Conception of *Episteme. Thomist* 39 (1975): 341–48.

Hall, Douglas C. *The Trinity: An Analysis of St. Thomas Aquinas' Expositio of the De Trinitate of Boethius.* Studien und Texte zur Geistesgeschichte des Mittelalters 33. Leiden: E. J. Brill, 1992.

Hamilton, Edith. *Mythology.* New York: New American Library, 1969.

Hamlyn, D. W. Aristotelian Epagoge. *Phronesis* 21 (1976): 167–84.

Hankey, Wayne J. The Structure of Aristotle's Logic and the Knowledge of God in the *Pars Prima* of the *Summa Theologiae* of Thomas Aquinas. In *Sprache und Erkenntnis im Mittelalter,* edited by Albert Zimmerman, 961–69. Berlin: Walter de Gruyter, 1981.

————. "Theoria Versus Poesis": Neoplatonism and Trinitarian Difference in Aquinas, John Milbank, Jean-Luc Marion and John Zizioulas. *Modern Theology* 15 (1999): 387–415.

Heidegger, Martin. *Der Satz Vom Grund.* Pfullingen: Verlag Gunther Neske, 1957.

————. *Metaphysische Anfangsgründe der Logik im Ausgang von Leibniz.* Gesamtausgabe 26. Frankfurt: V. Klostermann, 1978.

————. *The Metaphysical Foundations of Logic.* Translated by Michael Heim. Bloomington: Indiana University Press, 1984.

————. *Beiträge zur Philosophie (Vom Ereignis).* Edited by Friedrich-Wilhelm von Hermann. Gesamtausgabe 65. Frankfurt: Vittorio Klostermann, 1989.

————. *Principle of Reason.* Translated by Reginald Lilly. Bloomington: Indiana University Press, 1991.

————. *Besinnung.* Edited by Friedrich-Wilhelm von Hermann. Gesamtausgabe 66, 1997.

————. *Die Geschichte Des Seyns.* Edited by Peter Trawny. Gesamtausgabe 69. Frankfurt: Vittorio Klostermann, 1998.

Henry of Ghent. *Summae Quaestionum Ordinariarum.* St. Bonaventure: Franciscan Institute Publications, 1953.

————. *Quodlibet I.* Edited by Raymond Macken. Henrici de Gandavo Opera Omnia 5. Leuven: Leuven University Press, 1979.

Herrera, Robert A. Saint Thomas and Maimonides on the Tetragrammaton: The "Exodus" of Philosophy. *Modern Schoolman* 59 (1982): 179–93.

Herveus Natalis. *In Quatuor Libros Sententiarum Commentaria.* Paris: Dyonisius Moreau & Son, 1647.

Hesiod. *Theogony and Works and Days.* Translated by M. L. West. Oxford: Oxford University Press, 1988.

Hochstetter, Erich. Viator Mundi. Einige Bemerkungen zur Situation des Menschen bei Wilhelm von Ockham. *Franziskanische Studien* 32 (1950): 1–20.

Hoffmann, Fritz. Der Satz als Zeichen der Theologischen Aussage bei Holcot, Crathorn und Gregor von Rimini. In *Der Begriff der Repraesentatio im Mittelalter,* edited by Albert Zimmermann, 296–313. Miscellanea Mediaevalia 8. Berlin: Walter de Gruyter, 1871.

Honnefelder, Ludger. *Ens in Quantum Ens: Der Begriff des Seienden als Solchen als Gegenstand der Metaphysik nach der Lehre des Johannes Duns Scot.* Münster: Aschedendorff, 1979.

Humbrecht, Thierry-Dominique. La Theologie Negative Chez Saint Thomas d'Aquin. *Revue Thomiste* 94 (1994): 71–99.

Inagaki, B. Ryosuke. Res and Signum—On the Fundamental Ontological Presupposition of the Philosophy of William of Ockham. In *Philosophie im Mittelalter: Entwicklungslinien und Paradigmen,* edited by J. P. Beckmann et al., 301–11. Hamburg: Felix Meiner Verlag, 1987.

Jacobi Zabarella. *Opera Logica.* Frankfurt: Lazari Zetzneri, 1408.

Jean Buridan. *Compendium Totius Logicae.* Venice, 1499.

Joannes Glogoviensis. Jagiellonian University Library, 100r–155u. File: Cod. 25. Quationes Super Posteriora Analytica Aristotelis. Krakow.

————. *Quaestiones super Posteriora Analytica Aristotelis cum Commento J. Versoris.* Leipzig: W. Stoeckel, 1499.

Junghans, Helmar. *Ockham Im Lichte Der Neueren Forschung.* Hamburg: Lutherisches Verlagshaus, 1968.

Kahn, Charles. The Role of *nous* in the Cognition of First Principles in *Posterior Analytics* II 19. In *Aristotle on Science: The Posterior Analytics,* edited by Enrico Berti, 385–414. Padua: Editrice Antenore, 1981.

Kainz, Howard, P. *'Active and Passive Potency' in Thomistic Angelology.* The Hague: Martinus Nijhoff, 1972.

Kaufman, Matthias. Ockhams Logik und Erkenntnistheorie im Diskurs der Scholastic und in der Aktuellen Diskussion. *Philosophische Rundschau* 38 (1991): 318–28.

Kennedy, Leonard A. *Peter of Ailly and the Harvest of Fourteenth Century Philosophy.* Lewiston: Edwin Mellen Press, 1986.

————. *The Philosophy of Robert Holcot, Fourteenth Century Skeptic.* Lewiston: Edwin Mellen Press, 1993.

Klocker, Harry R. Ockham on the Cognoscibility of God. *Modern Schoolman* 35 (1958): 77–90.

Knasas, John. Ad Mentem Thomae: Does Natural Philosophy Prove God? *Proceedings of the Catholic Philosophical Association* 61 (1987): 209–20.

Köpf, Ulrich. *Die Anfänge der Theologischen Wissenschaftstheorie Im 13. Jahrhundert.* Tübingen: J. C. B. Mohr, 1974.

Kühn, Wilfried. *Das Prinzipienproblem in der Philosophie des Thomas von Aquin.* Amsterdam: Verlag B. R. Grüner, 1982.

Kusukawa, Sachiko. *The Transformation of Natural Philosophy: The Case of Philip Melancthon.* Cambridge: Cambridge University Press, 1995.

Lang, Albert. *Die Wege der Glaubensbegründung bei den Scholastikern des 14. Jahrhunderts.* Münster: Aschendorff, 1931.

Lee, Richard A. The Analogies of Being in St. Thomas Aquinas. *The Thomist* 58 (1994): 471–88.

————. Contingency from Necessity: The Late Medieval Grounding of Scientific Necessity. Ph.D. diss., Krakow, Poland, Jagiellonian University, 1995.

————. Peter Aureoli as Critic of Aquinas on the Subalternate Character of the Science of Theology. *Franciscan Studies* 55 (1998): 121–36.

Leff, Gordon. *William of Ockham: The Metamorphosis of Scholastic Discourse.* Manchester: Manchester University Press, 1975.

Lennox, J. G. Divide and Explain: The *Posterior Analytics* in Practice. In *Philosophical Issues in Aristotle's Biology,* edited by A. Gotthelf and J. Lennox, 90–119. Cambridge: Cambridge University Press, 1987.

Lesher, James A. The Meaning of *nous* in the *Posterior Analytics. Phronesis* 18 (1973): 44–68.

Libera, Alain de. *Penser Au Moyen Âge.* Paris: Éditions du Seuil, 1991.

Livesey, Steven J. Science and Theology in the Fourteenth Century: The Subalternate Sciences in Oxford Commentaries on the *Sentences. Synthese* 83 (1990): 273–92.

Lloyd, A. C. Necessity and Essence in the *Posterior Analytics.* In *Aristotle on Science: The Posterior Analytics,* edited by Enrico Berti, 157–71. Padua: Editrice Antenore, 1981.

Luna, Concetta. Theologie und Menschliche Wissenschaften in den *principia* des Aegidius Romanus. In *Scientia und Ars Im Hoch- und Spätmittelalter,* edited by

Ingrid Craemer-Rugenberg and Andreas Speer, 517–27. Miscellanea Mediaevalia 22/2. Berlin: Walter de Gruyter, 1994.

Lynch, Lawrence. The Doctrine of Divine Ideas and Illumination in Robert Grosseteste, Bishop of Lincoln. *Medieval Studies* 3 (1941): 163–73.

Lyttkens, Hampus. *The Analogy Between God and the World: An Investigation of Its Background and Interpretation of Its Use by Thomas of Aquino.* Uppsala: Almqvist & Wiksells Boktryckeri, 1952.

Magon, J. Ockham's Extreme Nominalism. *Thomist* 43 (1979): 414–49.

Maier, Anneliese. *Die Vorläufer Galileis.* Rome: Edizioni di Storia e Letteratura, 1949.

———. *On the Threshold of Exact Science: Selected Writings of Anneliese Maier on Late Medieval Natural Philosophy.* Edited and translated by Steven Sargent. Philadelphia: University of Pennsylvania Press, 1982.

Marrone, Steven P. *Truth and Scientific Knowledge in the Thought of Henry of Ghent.* Cambridge: The Medieval Academy of America, 1985.

Marsilius of Inghen. *Quaestiones super Quattuor Libros Sententiarum.* Edited by Manuel Santos Noya. Leiden: E. J. Brill, 2000.

Martin, Gottfried. *Wilhelm von Ockham: Untersuchungen Zur Ontologie der Ordnungen.* Berlin: Walter de Gruyter, 1949.

Maurer, Armand. Method in Ockham's Nominalism. *Monist* 61 (1978): 436–39.

McEvoy, James. *The Philosophy of Robert Grosseteste.* Oxford: Clarendon Press, 1982.

McInerny, Ralph. *The Logic of Analogy: An Introduction to St. Thomas.* The Hague: Martinus Nijhoff, 1961.

McKirahan, Richard D. *Principles and Proofs: Aristotle's Theory of Demonstrative Science.* Princeton: Princeton University Press, 1992.

Miethke, Jürgen. *Ockhams Weg Zur Social Philosophie.* Berlin: Walter de Gruyter, 1969.

Moody, Ernest. *The Logic of William of Ockham.* New York: Russell & Russell, 1965.

Murdoch, John. The Development of a Critical Temper: New Approaches and Modes of Analysis in 14th Century Philosophy, Science and Theology. *Medieval and Renaissance Studies* 7 (1975): 51–79.

Nemetz, Anthony A. Logic and the Division of the Sciences in Aristotle and St. Thomas Aquinas. *The Modern Schoolman* 33 (1956): 91–109.

Neumann, Siegfried. *Gegenstand und Methode der Theoretischen Wissenschaften nach Thomas von Aquin auf Grund der Expositio Super Librum Boethii De Trinitate.* Münster: Aschendorff, 1965.

Novak, Joseph Anthony. Aristotle on Method: Definition and Demonstration. Ph.D. diss., Notre Dame, Indiana, University of Notre Dame, 1977.

Nussbaum, Martha. *The Fragility of Goodness: Luck and Ethics in Greek Tragedy and Philosophy.* Cambridge: Cambridge University Press, 1986.

Oakley, Francis. Pierre d'Ailly and the Absolute Power of God: Another Note on the Theology of Nominalism. *Harvard Theological Review* (1963): 59–73.

———. *Omnipotence, Covenant and Order. An Excursion in the History of Ideas from Abelard to Leibniz.* Ithaca: Cornell University Press, 1984.

Oberman, Heiko A. *Forerunners of the Reformation: The Shape of Late Medieval Thought.* London: P. Luttworth, 1967.

————. *The Dawn of the Reformation*. Edinburgh: T & T Clark, 1986.

————. *Via Antiqua* and *Via Moderna*: Late Medieval Prolegomena to Early Reformation Thought. In *From Ockham to Wyclif*, edited by Anne Hudson and Michael Wilks, 445–63. Oxford: Basil Blackwell, 1987.

O'Connor, Edward. The Scientific Character of Theology According to Scotus. In *De Doctrina Ioannis Duns Scoti*, edited by Commisio Scotistica, 3–50. Rome: Commisio Scotistica, 1968.

Owens, Joseph. Common Nature: A Point of Comparison Between Scotistic and Thomistic Metaphysics. *Medaeval Studies* 19 (1957): 1–14.

Patzig, Günther. Erkenntnisgründe, Realgründe und Erklärungen. In *Aristotle on Science: The Posterior Analytics*, edited by Enrico Berti, 141–56. Padua: Editrice Antenore, 1981.

Paulus Venetus. *Expositio in Libros Posteriorum Anlyticorum Aristotelis*. Venice, 1481.

Pegis, Anton. Concerning William of Ockham. *Traditio* 2 (1944): 465–80.

Pernoud, M. A. Innovation in William of Ockham's References to the Potentia Dei. *Antonianum* 45 (1970): 65–97.

————. The Theory of the Potentia Dei According to Aquinas, Scotus and Ockham. *Antonianum* 47 (1972): 69–95.

Peter Aureoli. *Scriptum Super Primum Sententiarum*. Edited by Eligius Buytaert, O.F.M. St. Bonaventure: Franciscan Institute Publications, 1956.

Petrus de Aquila. *Quaestiones in IV Libros Sententiarum*. Speyer: 1480.

Porphyry. *Isagoge*. Translated by Edward W. Warren. Mediaeval Sources in Translation 16. Toronto: Pontifical Institute of Medieval Studies, 1975.

Prouvost, Gery. La Tension Irresolue. *Revue Thomiste* 98 (1998): 95–102.

Randi, Eugenio. A Scotist Way of Distinguishing Between God's Absolute and Ordained Power. In *From Ockham to Wyclif*, edited by Anne Hudson and Michael Wilks, 43–50. Oxford: Basil Blackwell, 1987.

Rijk, L. M. de. Questio de Ideis. Some Notes on an Important Chapter of Platonism. In *Kephalaion: Studies in Greek Philosophy and Its Continuation*, edited by J. Mansfeld and L. M. de Rijk, 204–13. Assen: Van Gorcum, 1975.

————. War Ockham ein Antimetaphysiker? Eine Semantische Betrachtung. In *Philosophie im Mittelalter*, edited by Jan P. Beckmann, 313–28. Hamburg: Felix Meiner Verlag, 1987.

Robertus Grosseteste. *Commentarius in Posteriorum Analyticorum Libros*. Edited by Pietro Rossi. Firenze: Leo S. Olschki, 1981.

Robertus Holkot. *Quaestiones super IV Libros Sententiarum*. Lyon: Joannes Trechsel, 1497.

Rocca, Gregory. The Distinction Between *res Significata* and *modus Significandi* in Aquinas's Theological Epistemology. *Thomist* (1991): 173–97.

Roger Bacon. An Unedited Part of Roger Bacon's "Opus Maius": "De Signis." *Traditio* 34 (1978): 75–136.

Rudavsky, Tamar M. The Doctrine of Individuation in Duns Scotus. *Franziskanische Studien* 59 (1977): 320–77.

————. The Doctrine of Individuation in Duns Scotus. *Franziskanische Studien* 60 (1980): 62–83.

Schmidt, Werner. *Theorie der Induktion.* München: Wilhelm Fink Verlag, 1974.

Schultess, Peter. "Significatio" Im Rahmen der Metaphysik (Kritik) Ockhams. *Vivarium* 29 (1991): 104–28.

———. *Sein, Signifikation und Erkenntnis bei Wilhelm von Ockham.* Berlin: Akademie Verlag, 1992.

Schürmann, Reiner. *Heidegger on Being and Acting: From Principles to Anarchy.* Translated by Christine Marie Gros. Bloomington: Indiana University Press, 1987.

Science. Edited by Jonathan Barnes, Malcolm Schofield, and Richard Sorabji. Articles on Aristotle 1. London: Duckworth, 1975.

Scott, T. K. Ockham on Evidence, Necessity and Intuition. *Journal of the History of Philosophy* 9 (1971): 15–41.

Serene, Eileen. Robert Grosseteste on Induction and Demonstrative Science. *Synthese* 40 (1979): 97–115.

Smith, Robin. The Syllogism in *Posterior Analytics* I. *Archiv für Geschichte der Philosophie* 64 (1982): 113–35.

———. Immediate Propositions and Aristotle's Proof Theory. *Ancient Philosophy* 6 (1986): 47–68.

Streveler, Paul A. Ockham and His Critics on: Intuitive Cognition. *Franciscan Studies* 35 (1975): 223–36.

Tachau, Katherine. *Vision and Certitude in the Age of Ockham: Optics, Epistemology and the Foundations of Semantics 1250–1345.* Studien und Texte zur Geistesgeschichte des Mittelalters XXII. Leiden: E. J. Brill, 1988.

Te Velde, Rudi A. *Participation and Substantiality in Thomas Aquinas.* Leiden: E. J. Brill, 1995.

Thijssen, J. M. M. H. Once Again the Ockhamist Statutes of 1339 and 1340: Some New Perspectives. *Vivarium* 28 (1990): 136–67.

Thomas Angelicus. *Liber Propugnatoris Super Primum Sententiarum Contra Johannem Scotum.* Venice, 1523.

Thomas Aquinas. *Sancti Thomae de Aquino Opera Omnia Iussu Leonis XIII P.M. Edita.* Edited by Commissio Leonina. 50 vols. Rome: Commissio Leonina, 1882-.

———. *In Aristotelis Libros Posteriorum Analyticorum Expositio.* Edited by Raymund M. Spiazzi. Turin: Marietti, 1955.

———. *Liber de Veritate Catholicae Fidei Contra Errores Infidelium, Seu "Summa Contra Gentiles."* Edited by Ceslai Pera, Petro Marc, and Petro Caramello. Rome: Marietti, 1961.

———. *Expositio Super Librum Boethii de Trinitate.* Edited by Bruno Decker. Studien und Texte zur Geistesgechichte des Mittelalters IV. Leiden: E. J. Brill, 1965.

———. *S. Thomae Aquinatis Opera Omnia: Ut Sunt in Indice Thomistico.* Edited by Roberto Busa. 7 vols. Stuttgart-Bad Cannstatt: Frommann-Holzboog, 1980.

———. *St. Thomas Aquinas. Faith, Reason and Theology: Question I-IV of His Commentary on the De Trinitate of Boethius.* Edited and translated by Armand Maurer. Toronto: Pontifical Institute of Medieval Studies, 1987.

———. *Expositio Libri Peryermenias.* Vol. I pt. 1 of *Sancti Thomae de Aquino Opera Omnia Iussu Leonis XIII P.M. Edita.* Edited by René-Antoine Gauthier O.P. Paris: J. Vrin, 1989.

————. *Expositio Libri Posteriorum*. Vol. I, pt. 2 of *Sancti Thomae de Aquino Opera Omnia Iussu Leonis XIII P.M. Edita*. Edited by René-Antoine Gauthier O.P. Paris: J.Vrin, 1989.

————. *Thomae Aquinatis Opera Omnia [Computer File]: Cum Hypertextibus in CD-ROM*. Edited by Roberto Busa. Milan: Editoria Elettronica Editel, 1992.

Vier, P. *Evidence and Its Function According to John Duns Scotus*. St. Bonaventure: Franciscan Institute Publications, 1951.

Vossenkühl, Wilhelm. Ockham on the Cognition of Non-Existents. *Franciscan Studies* 45 (1985): 33–45.

Wallace, William. Buridan, Ockham, Aquinas: Science in the Middle Ages. *Thomist* 40 (1976): 475–83.

Webering, Damascene, O.F.M. *Theory of Demonstration According to William of Ockham*. St. Bonaventure: Franciscan Institute Publications, 1953.

Weisheipl, James A. The Concept of Nature: Avicenna and Aquinas. In *Thomistic Papers I*, edited by Victor Brezik, 65–81. Houston: Center for Thomistic Studies, 1980.

Wengert, R. G. The Sources of Intuitive Cognition in William of Ockham. *Franciscan Studies* 41 (1981): 415–44.

Wilcox, Donald J. *In Search of God and Self: Renaissance and Reformation Thought*. Boston: Houghton Mifflin Company, 1975.

Wilcox, John R. Our Knowledge of God in "Summa Theologiae, Prima Pars, Quaestiones 3–6": Positive or Negative? *American Catholic Philosophical Quarterly* 75, Supp. (1998): 201–11.

William of Ockham. *Scriptum in Librum Primum Sententiarum Ordinatio*. Prologus et Distinctio I. Edited by Gedeon Gál, O.F.M., and Stephen Brown, O.F.M. Guillelmi de Ockham Opera Philosophica et Theologica OT I. St. Bonaventure: Franciscan Institute Publications, 1967.

————. *Scriptum in Librum Primum Sententiarum Ordinatio*. Distinctiones II et III. Edited by Girard Etzkorn. Guillelmi de Ockham Opera Philosophica et Theologica OT II. St. Bonaventure: Franciscan Institute Publications, 1970.

————. *Summa Logicae*. Edited by Philotheus Boehner, O.F.M., Gedeon Gál, O.F.M., and Stephan Brown. Guillemi de Ockham Opera Philosophica et Theologica OP I. St. Bonaventure: Franciscan Institute Publications, 1974.

————. *Scriptum in Librum Primum Sententiarum Ordinatio*. Distinctiones IV-XVIII. Edited by G. Etzkorn. Guillelmi de Ockham Opera Philosophica et Theologica OT III. St. Bonaventure: Franciscan Institute Publications, 1977.

————. *Expositio in Librum Praedicamentorum Aristotelis*. Edited by Gedeon Gál, 135–339. Guillelmi de Ockham Opera Philosophica et Theologica OP II. St. Bonaventure: Franciscan Institute Publications, 1978.

————. *Quodlibeta Septem*. Edited by J. C. Wey. Guillelmi de Ockham Opera Philosophica et Theologica OT IX. St. Bonaventure: Franciscan Institute Publications, 1980.

Wilpert, Paul. *Das Problem der Wahrheitssicherung bei Thomas von Aquin: Ein Beitrag zur Geschichte des Evidenzproblems*. Beiträge zur Geschichte der Philosophie und Theologie des Mittelalters XXX, heft 3. Münster: Aschendorff, 1931.

Wolter O.F.M., Allan B. *The Transcendentals and Their Function in the Metaphysics of Duns Scotus.* St. Bonaventure: The Franciscan Institute, 1946.

————. Ockham and the Textbook: On the Origin of Possibility. *Franziskanische Studien* 32 (1950): 70–96.

Wood, Rega. Intuitive Cognition and Divine Omnipotence: Ockham in Four-teenth-Century Perspective. In *From Ockham to Wyclif,* edited by Anne Hudson and Michael Wilks, 51–61. London: Basil Blackwell, 1987.

————. Ockham on Essentially-Ordered Causes: Logic Misapplied. In *Die Gegenwart Ockhams,* edited by Wilhelm Vossenkuhl and Rolf Schönberger, 25–50. Weinheim: VCH-Verlagsgesellschaft, 1990.

Wright, John H., S.J. *The Order of the Universe in the Theology of St. Thomas Aquinas.* Analecta Gregoriana 89. Rome: Gregorian University Press, 1957.

INDEX

Bold typeface indicates items that receive sustained treatment in the text.

Abstractive cognition, 86
  intuitive cognition as distinct from, 85
Abstractive knowledge, 71, 86, 90, 92, 148
  as basis for science, 65, 70, 71
  evident knowledge and, 93
  intuitive knowledge as cause of, 93, 141
  intuitive knowledge as distinct from, 66, 71, 86, 93
  of singulars (in Scotus), **66–68**
  of God (in Ockham), 93
  of God (in Scotus), 92
Adorno, Theodor, 119, 122, 123, 125, 157
Anaxagoras, 2, 5
Apprehension
  abstractive (in Ockham), 148, **85–87**
  in Ockham, **81–87**
  intellectual, 82, 140
  intuitive (in Ockham), **85–87**
  sensory, 82, 140
Aquinas, Thomas, 6, 33, 34, 35, 36, 39, 42, 45, 46, 59, 127, 128, 130
  apprehension in, **82–85**
  intellectus and, 47
  judgment in, **82–85**
  negative theology and, 54–55
  Ockham's rejection of, 94
  per-se predication and, **33–36**

negative and analogical character of theology and, **54–57**
theology as subalternate science and, **33–57**
unicity of God and world in, 59
Aristotelian science
  divine ideas and, **17–32**
  intuitive knowledge and, 103
  theological propositions and, **13–15**
Aristotle, 3, 4, 5, 7, 8, 9, 10, 11, 12, 13, 14, 26, 29, 31, 60, 65, 121
  See also Categories, De Caelo, Metaphysics, Peri Hermeneias, Posterior Analytics, Prior Analytics

Beatific vision, 44, 57, 79, 83, 92, 127, 134, 147
  as natural end of rational being, 45
  certainty of, 45
  character of, 44, 52
  excessive character of, 52
  intellectus and, 52
  intuitive cognition and, 65
  necessity of, 45
  theological propositions and, 93
Being
  concept of, 63
  identification of knowing with, 120
Blessed
  requirement for theology of knowledge of the, 44
  science of the, 40, 43, **44–45**

Categories, 35, 132
Cognitio intuitiva, 64